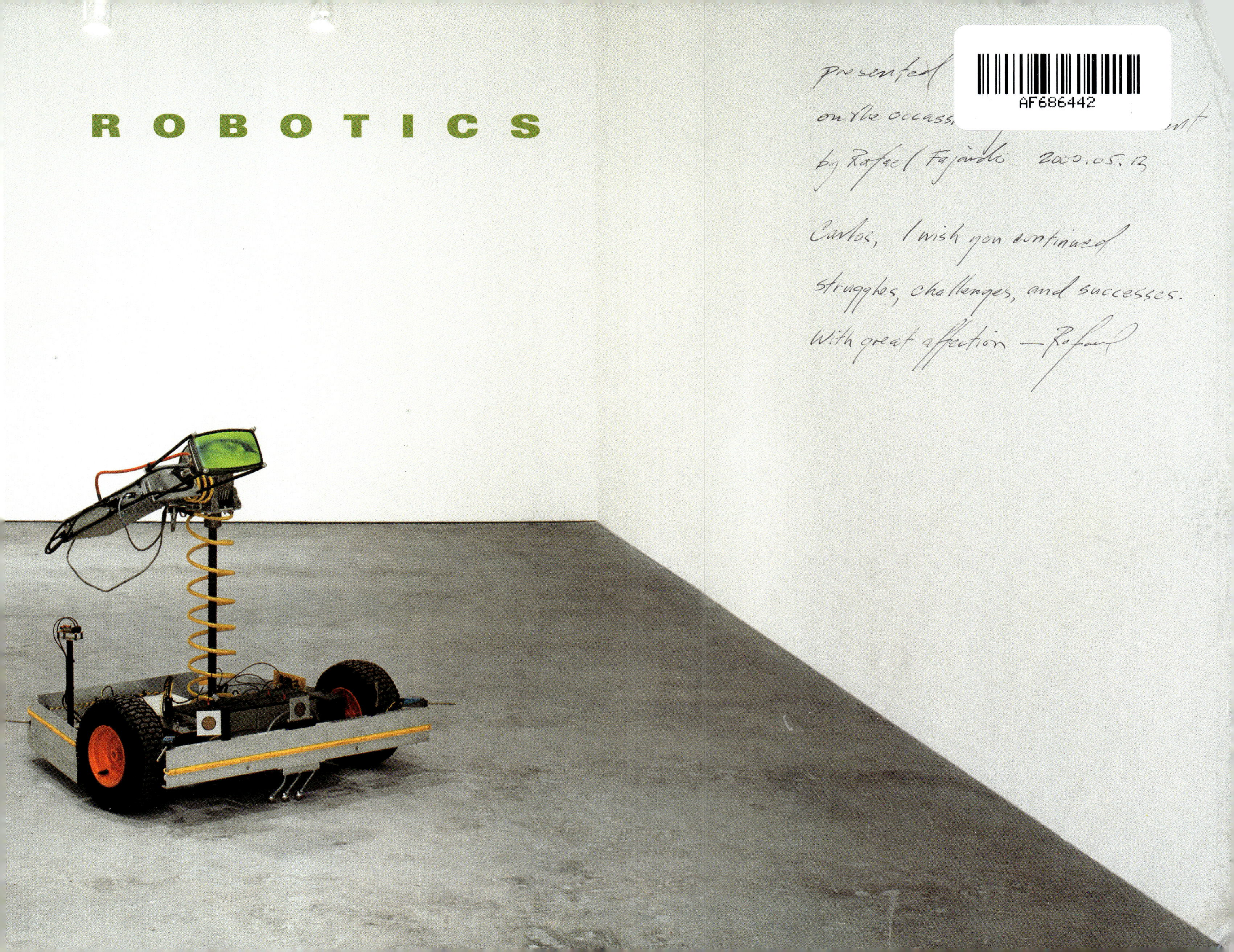

ROBOTICS

presented [illegible]
on the occasi[illegible] [illegible]ent
by Rafael Fajardo 2000.05.12

Carlos, I wish you continued
struggles, challenges, and successes.
With great affection —Rafael

ROVER
1998
detail & installation view
Computers, software,
electronics, aluminum,
motors
approx. 44 x 38 x 41 inches

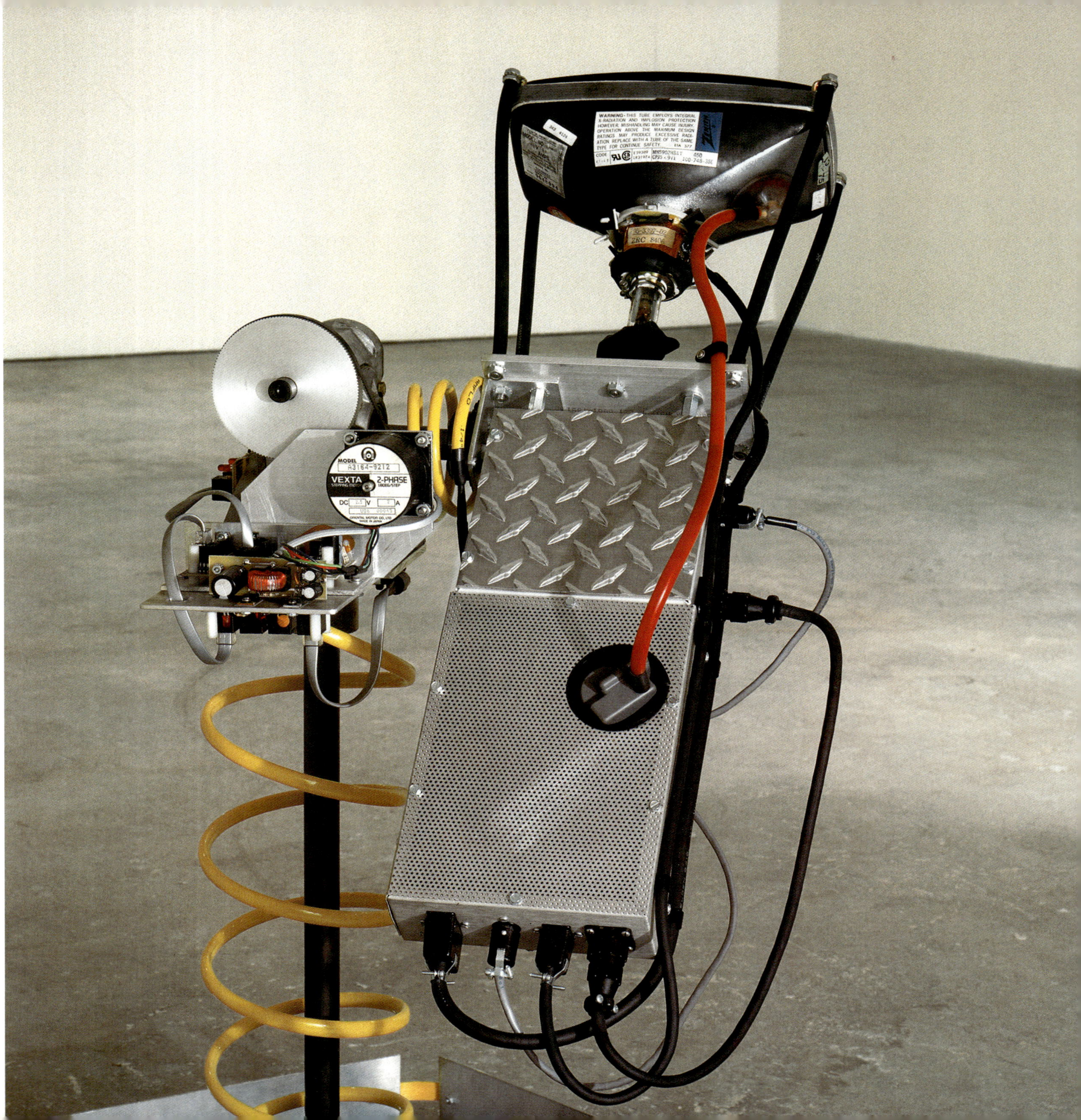

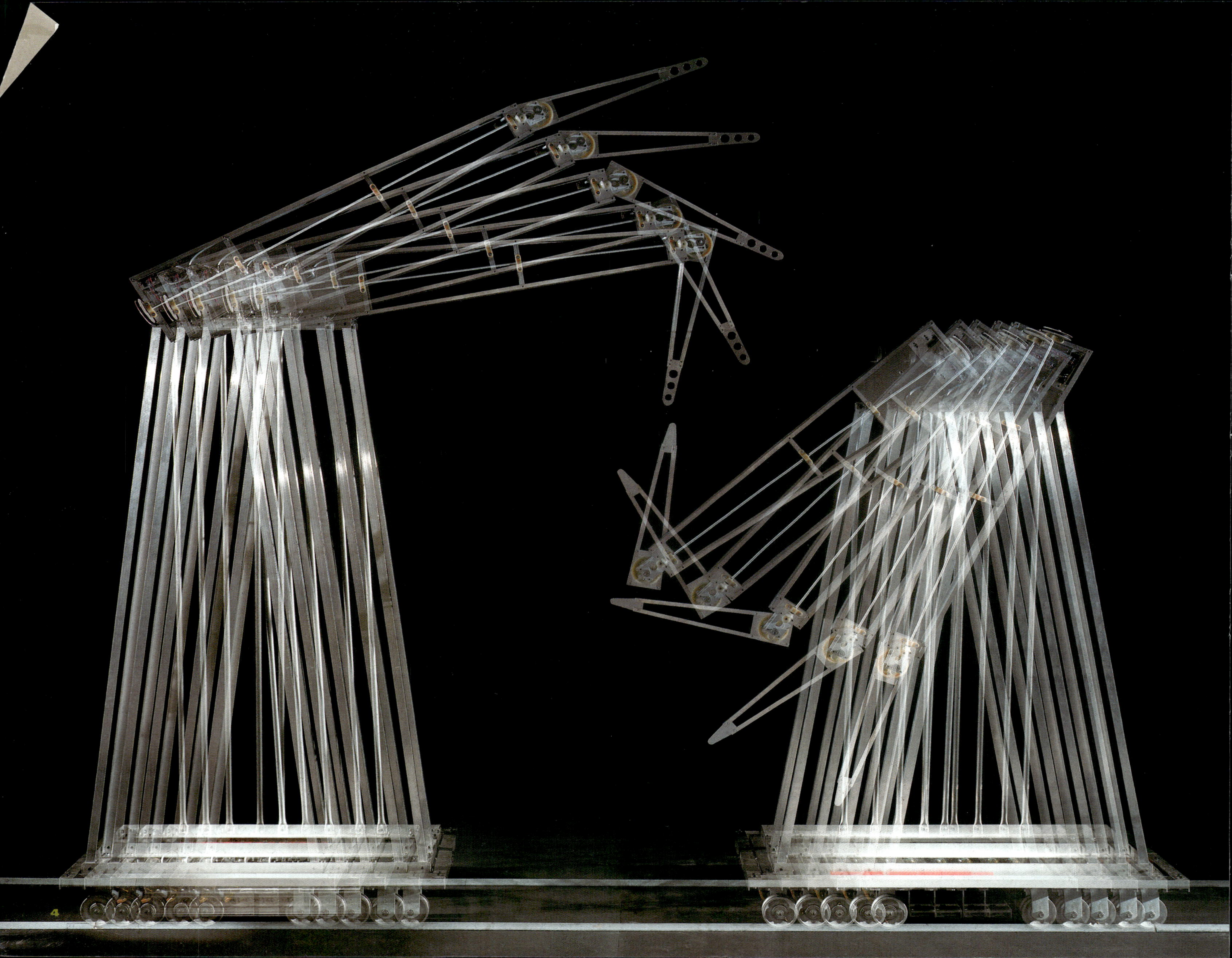

ONE TRACK MINDS
1998
Computers, software
aluminum, motors
Track: 40 feet long
Two units:
11 x 4 x 2 feet,
13 x 3 x 2 feet

If the story of the 19th century was the machine (the extension of human control over the environment by mechanical means) and the story of the 20th century is electronics (radical extension of that control by harnessing the electron), then the story of the 21st century may be best described by Thomas Pynchon as a time "when the curves of research and development in artificial intelligence, molecular biology and robotics all converge."[1] Such a situation could make possible the creation of an artificial life in which distinctions between the synthetic and the organic would disappear. Artist Alan Rath charts the territory of this imminent future. SITE Santa Fe is pleased to organize *Alan Rath: Robotics* as well as commission two new works—*Rover* (1998) and *Inchworm* (1998)—specifically for the exhibition.

Rath manipulates electronics as both formal and metaphorical elements to create inventive sculptures that comment on the symbiotic relationship between humans and machines and that relationship's implications on our present and future. While Rath's education and background are solidly grounded in engineering, his work transgresses science to become art. He employs eviscerated, deliberately physical machine components and anatomically integrated electronics to create his sculptures.

For the exhibition, Rath creates an artificial ecosystem with mechanized forms affected by arbitrary stimuli. This environment is populated by a variety of electromechanical species that vary from simple mobile "plants" to complex mobile "animals." Members of this species have varying degrees of perceptual ability: the simplest have no knowledge of the environment; others are able to sense their companion species; and the most complex interact with visitors to the exhibition.

Rath's robotic sculptures can be seen as icons that represent both continuing human dilemmas and nascent cyberquestions. They ask us to face questions about the way we live as we confront competition and exchange in a technological society. These issues have a place in both art and in the larger world, especially now, as we approach the beginning of the 21st century.

This exhibition would not have been possible without the generous support of Toby Devan Lewis and the following organizations: Elizabeth Firestone Graham Foundation, The Max and Victoria Dreyfus Foundation, Inc., The Shifting Foundation, Smart Art Press, and The Vicki and Kent Logan Family Foundation. I would also like to thank Alan Rath and the staff of SITE Santa Fe, as well as Elizabeth Ferrer, director, at the Austin Museum of Art and Robert Knight, director, and Debora Hopkins, curator, at the Scottsdale Museum of Contemporary Art for participating in the tour of the exhibition. Special thanks are due to Cheryl Haines for her commitment to this project.

LOUIS GRACHOS
Director & Curator

1 "Is It O.K. to Be a Luddite?," Thomas Pynchon, *The New York Times Book Review*, Oct. 28, 1984, 40–41.

CALIFORNIA LANDSCAPE

1998

installation view & detail

Computers, software, electronics, aluminum, motors

Two units:

12 x 25 x 12 feet overall

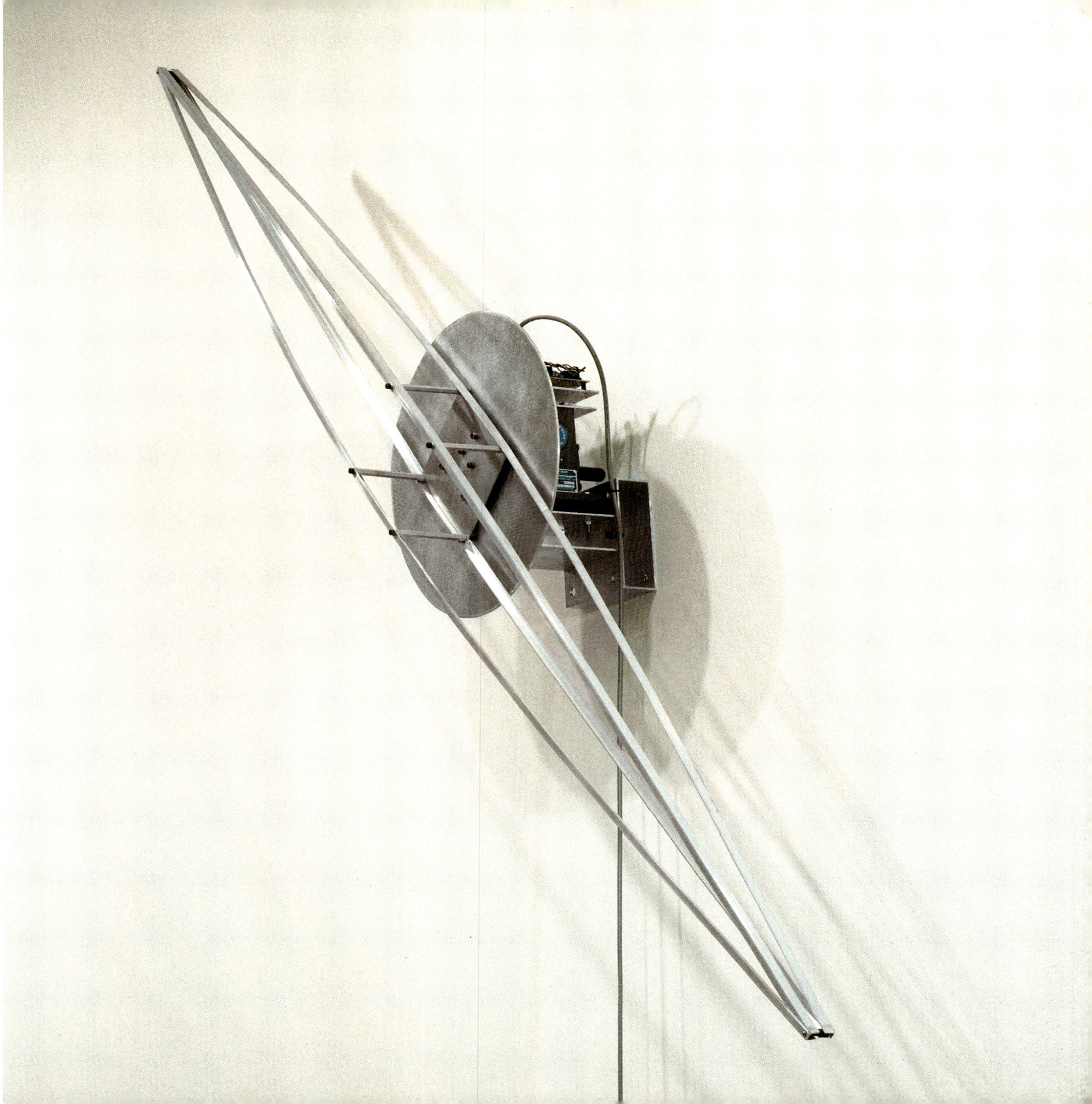

CALIFORNIA LANDSCAPE

1998

installation view

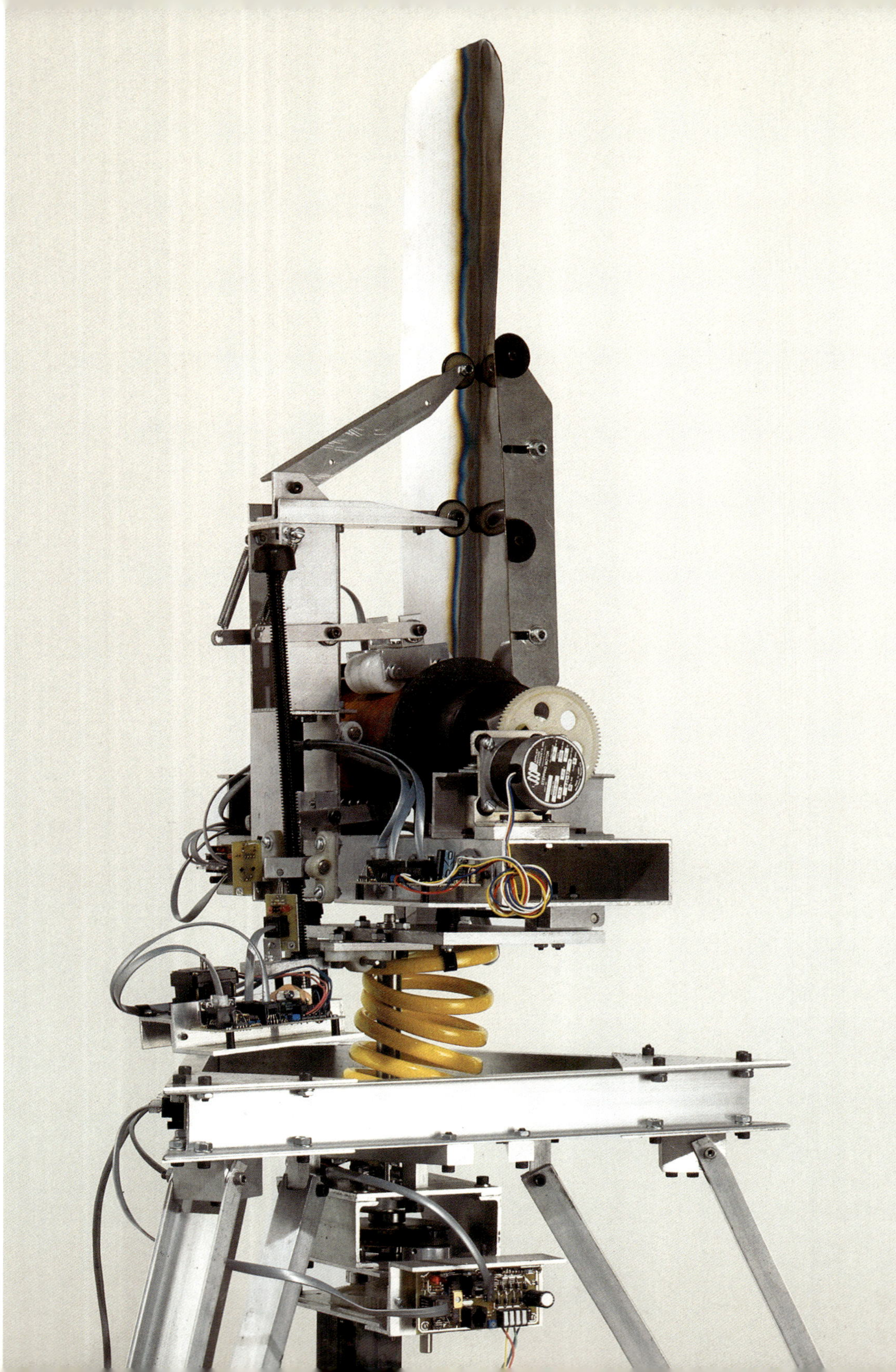

FRIENDS AND ACQUAINTANCES
1998
detail & installation view
Computers, software, electronics, aluminum, motors
Five units:
12 x 33 x 14 feet overall

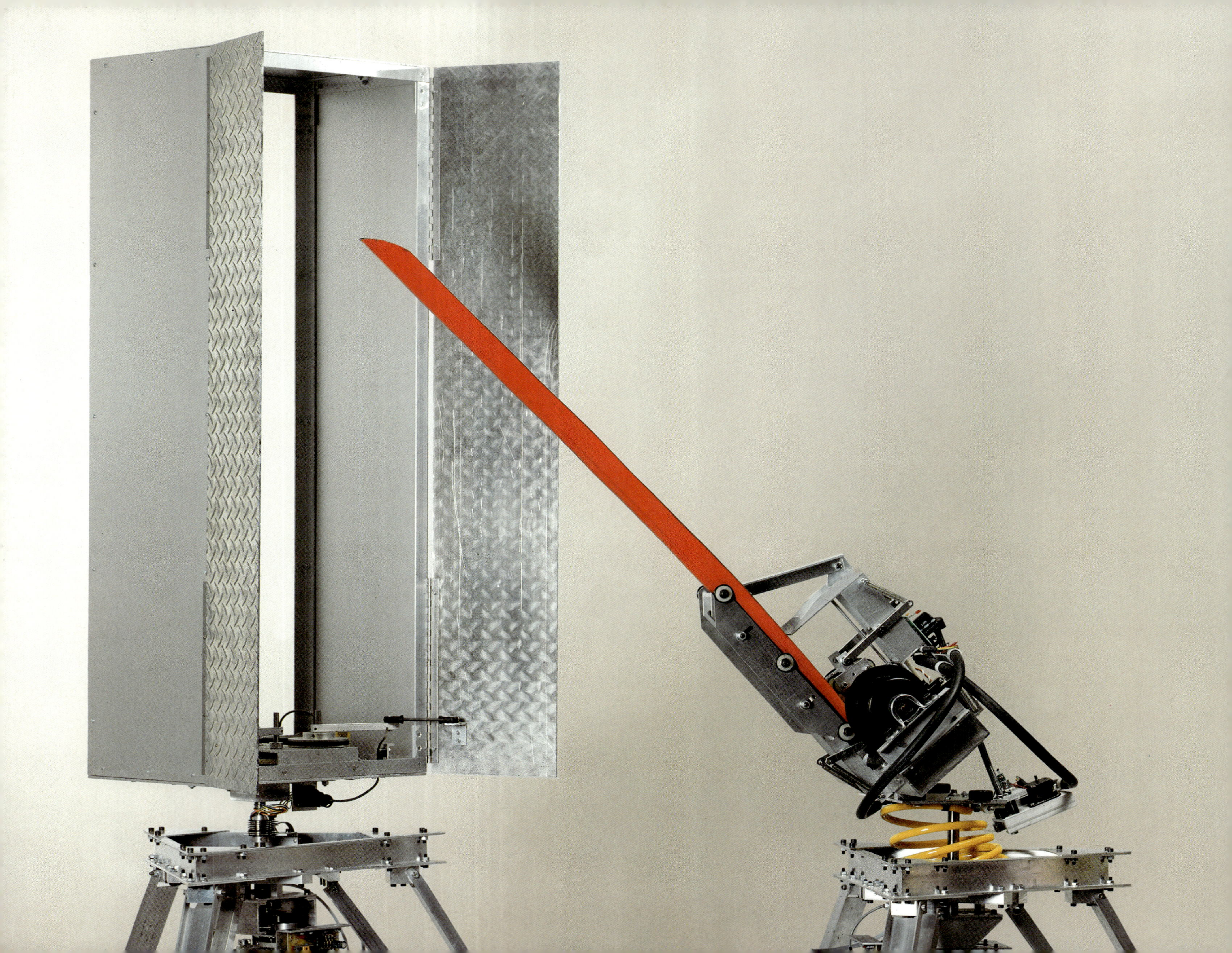

FRIENDS AND

ACQUAINTANCES

1998

details

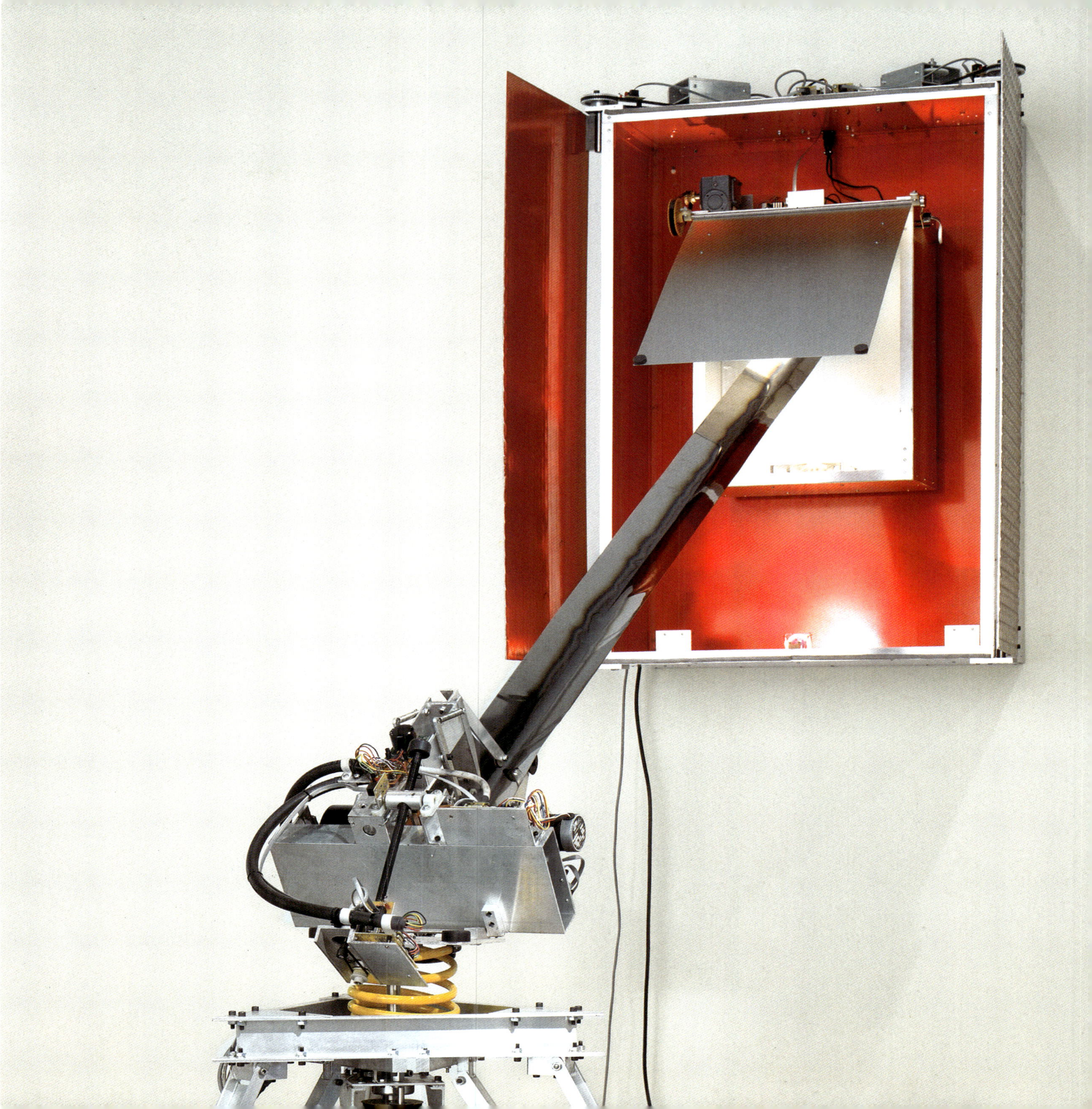

ROBOT DANCE

1995

Computer, software, electronics, aluminum, motors

Two units:

10 x 20 x 3 feet overall

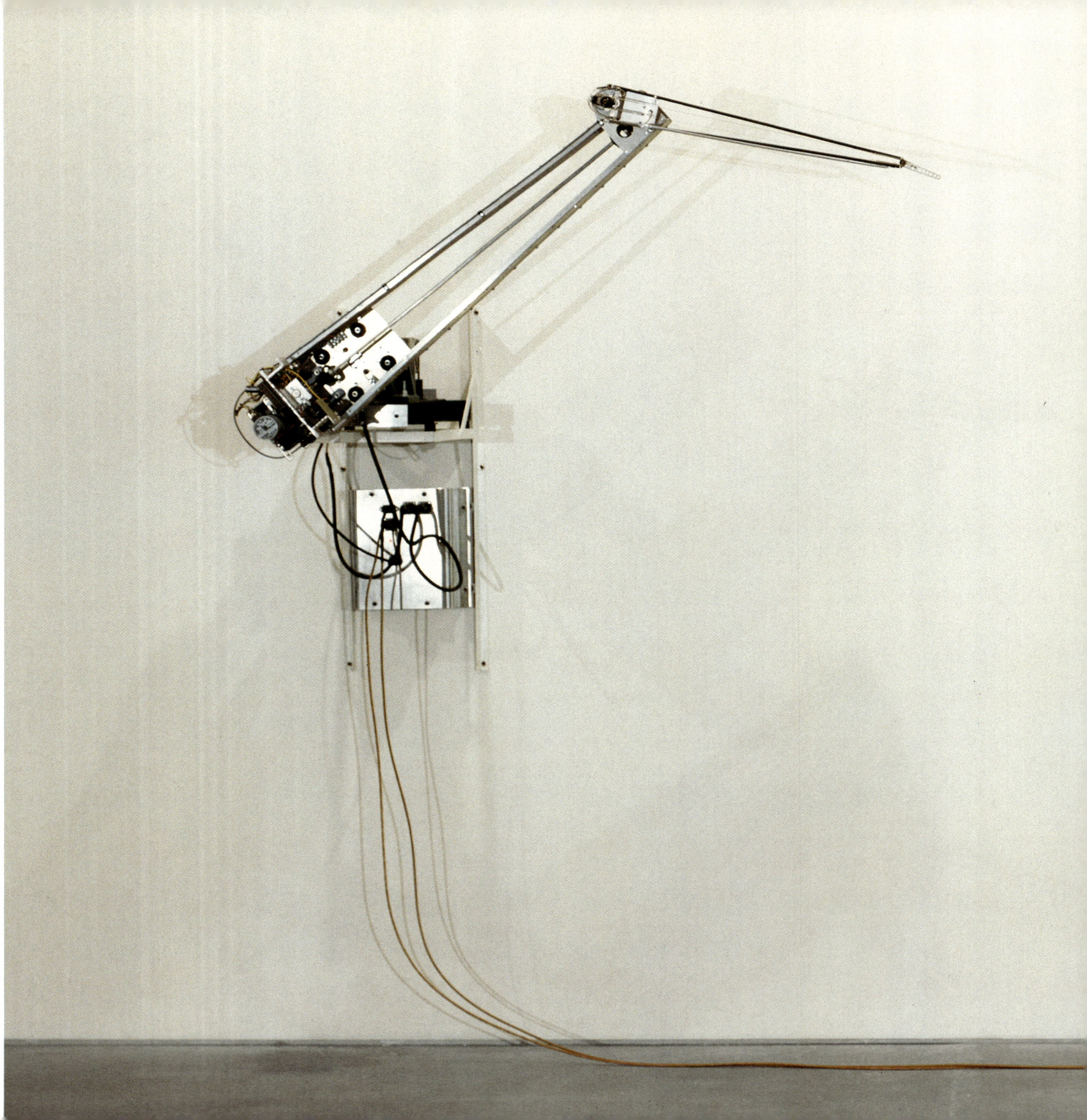

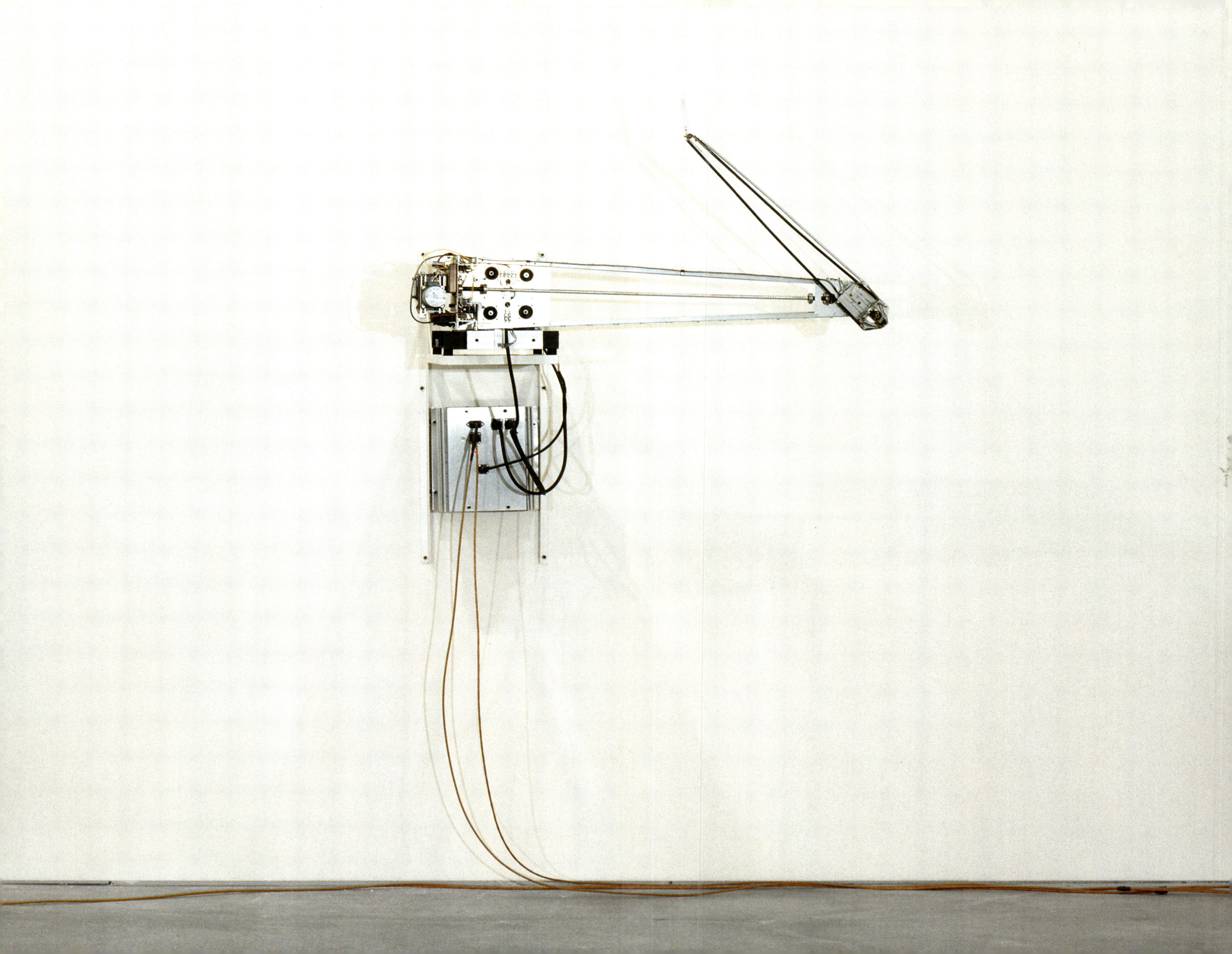

ROBOT DANCE

1995

details

I am intrigued and delighted by the œuvre of Alan Rath, an artist whose elegant machines are the subject of this exhibition. As an artist who studied physics and electrical engineering at the Massachusetts Institute of Technology, Rath is able to design and build his computerized sculptures all by himself, and they appeal to the scientist in me as well as the art lover.

Some are programmed to move in response to the presence of people nearby. Others interact in pairs, perhaps bowing to each other like dancers in an 18th-century ballroom. Even though they have the look of machines, they remind us, through their motions, of living organisms.

A Rath device has its food, in the form of electrical energy, supplied through a fixed plate in the floor. When, in the course of its wanderings, the sculpture returns to its feeding pad, it can take up energy and be ready for further adventures. Watching as it just misses the pad time after time and then finally reaches it, we feel the kind of sympathy we would normally extend to a living creature.

One can think of these creations as robots, but that word (Czech for worker) seems to imply that they serve us in some banal capacity. Instead they serve to entertain us, stimulate our thoughts, play on our feelings, and excite our imaginations. Some utilize pictures or sound or both, while others just gesture in elaborate ways and move about. All are programmed to follow a course that is partly deterministic but affected by random perturbations. In that way they illustrate the prevalence, in virtually all kinds of systems, of a combination of regular patterns and randomness.

Even the universe as a whole exhibits such a combination. It obeys the fundamental laws of physics. Those consist of the long-sought unified quantum field theory (which may finally have been discovered in the form of "superstring" or "M" theory) and the initial condition of the universe, characterizing its state at the beginning of its expansion around 13 billion years ago. But those laws are not fully deterministic—they are probabilistic and yield the probabilities for alternative histories of the universe. The history we experience is codetermined by the laws and outcomes of an inconceivably long sequence of chance events (or "accidents"). The universe is thus a prototype of systems exhibiting regularity supplemented by randomness.

Actually, in constructing his "random" input, Alan Rath cheats a bit, just as most scientists and mathematicians do today. Instead of using numbers generated by some truly random physical effect — years ago, noise from a vacuum tube was popular — he employs a deterministic computer program that produces a messy, random-looking sequence of numbers. Such an input is called pseudo-random. Fortunately, the viewer cannot tell the difference.

So much of our experience involves systems that combine the regular and the random. As a result, Alan Rath's sculptures invoke thoughts and feelings about an astonishing variety of situations. Think of the courtship ceremony of a species of bird or mammal. It follows a pattern all right, but a great many details can vary from one couple to another and also from one occasion to another, even for the same couple. In cooking, we may follow a recipe, but the details are not fixed — the exact proportions of the ingredients, the cooking times, the temperature at which the food is served and eaten. When we listen to a pianist playing the same piece on two different occasions, the performances are very similar, but we can pick out some differences nevertheless. When a factory produces parts for a car, each part is made to strict manufacturing standards, but within the specifications there is always some room for random variation.

Alan Rath's splendid creations are a powerful reminder of the ubiquity of regularity and chance variation.

MURRAY GELL-MANN

MURRAY GELL-MANN is Distinguished Fellow and Co-Chair of the Science Board at the Santa Fe Institute, and a member of the President's Council of Advisors on Science and Technology.

INCHWORM

1998

Computers, software, electronics, aluminum, motors

dimensions variable, approx. 73 inches high

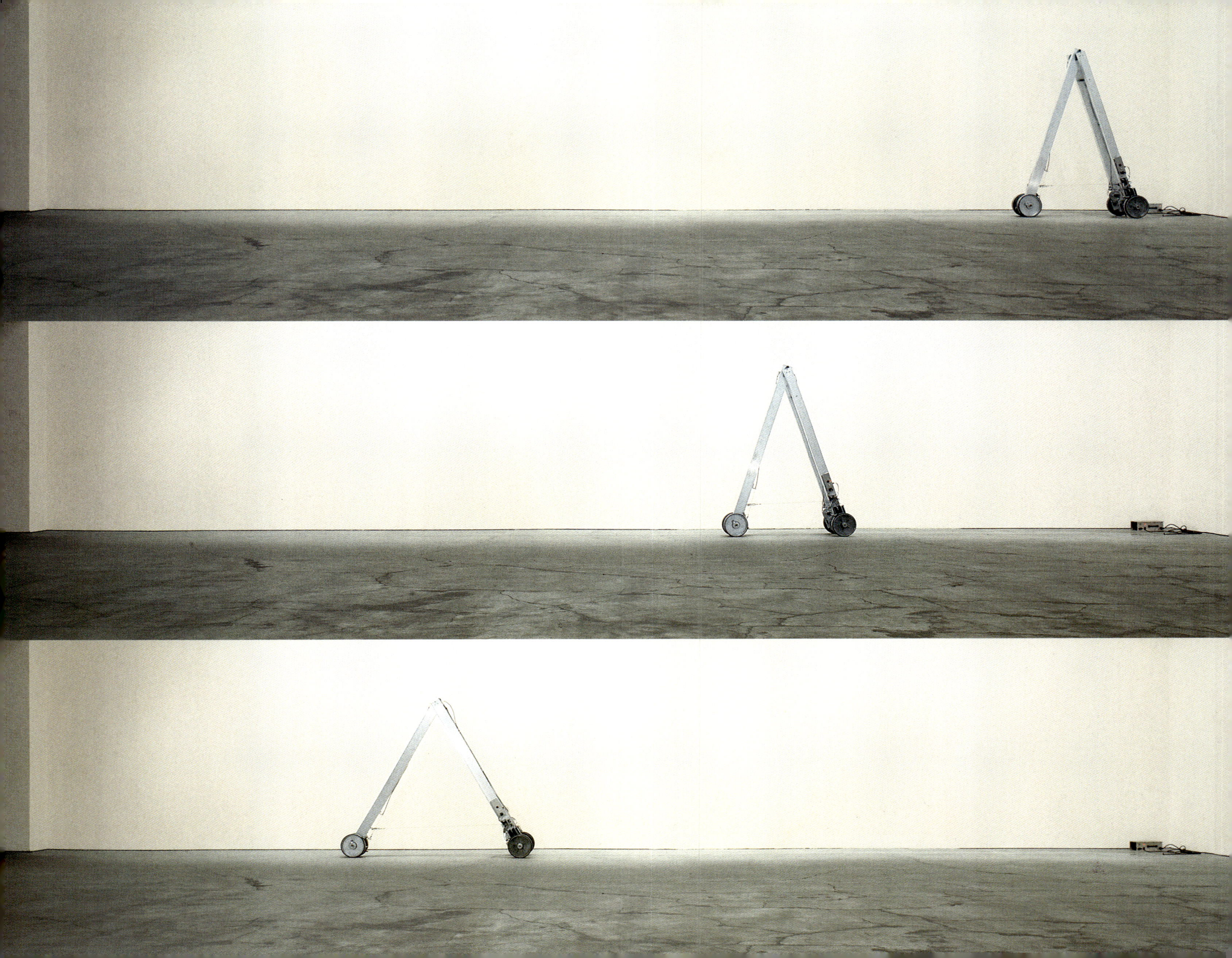

INCHWORM

1998

details

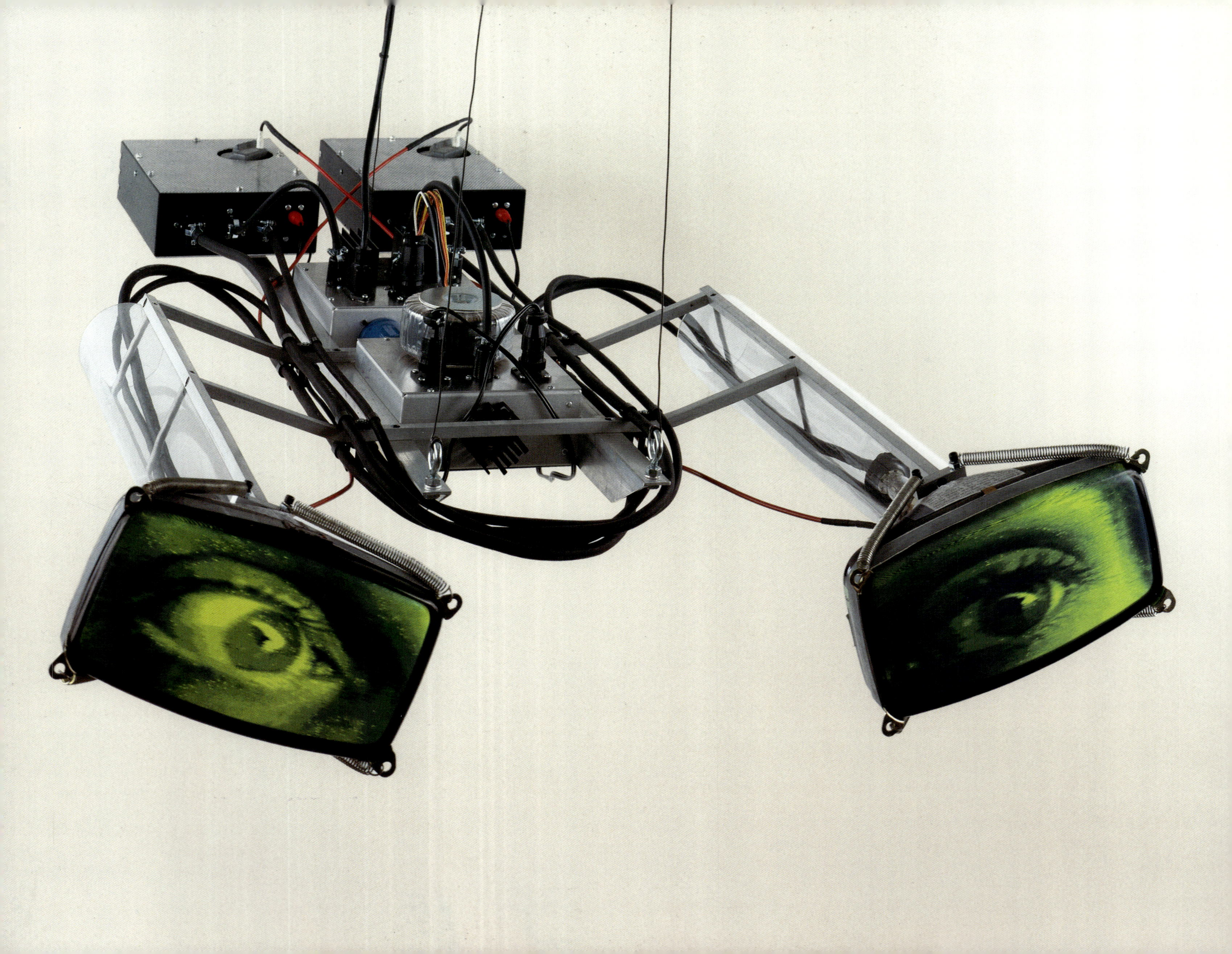

SOAR EYES

1994

Aluminum, acrylic, electronics, cathode ray tubes

9 x 32 x 45 inches

FIVE ON THE WALL

1998

Computers, software, electronics, aluminum, motors

Five units:

16 x 55 x 2 feet overall

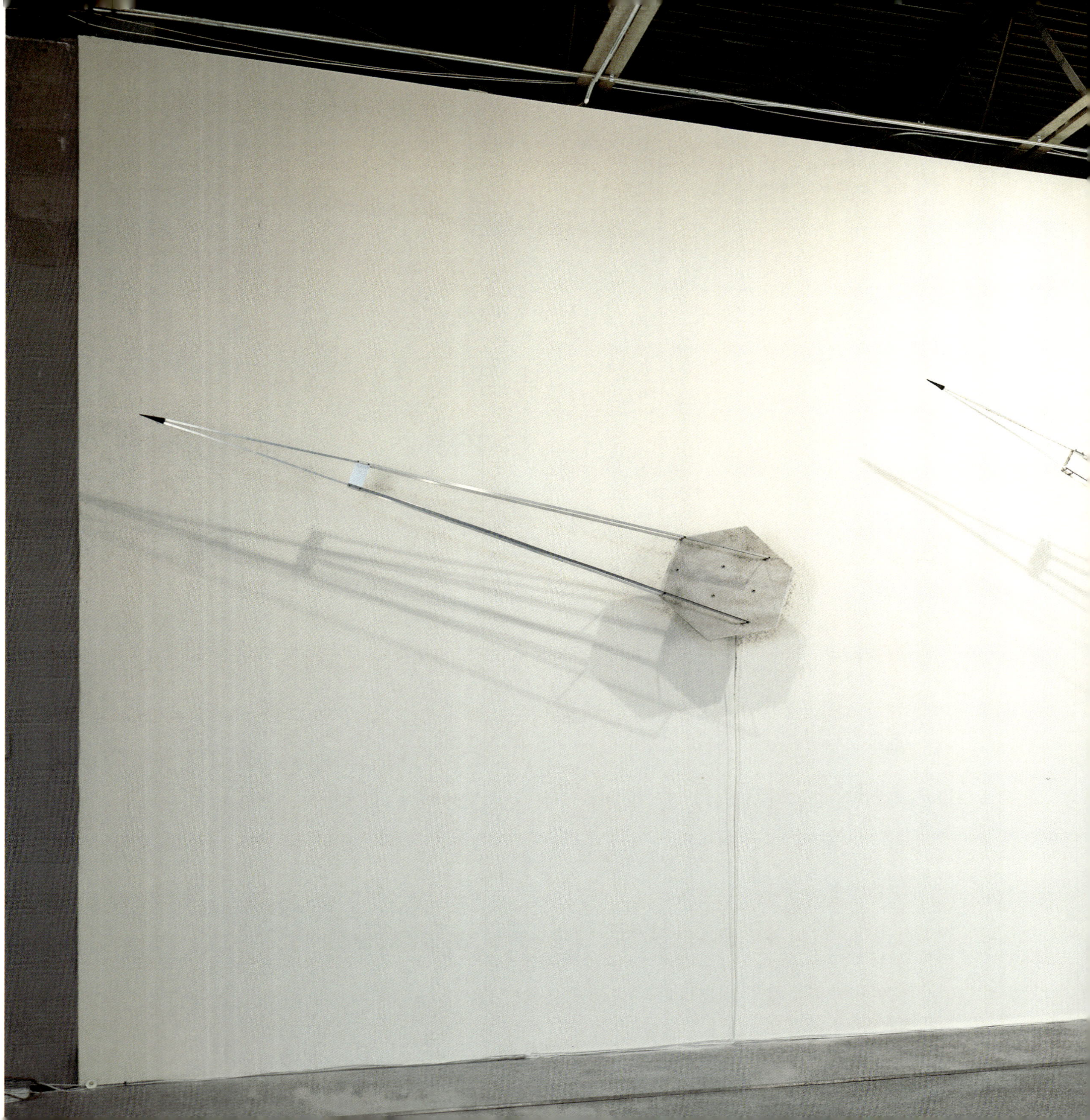

BOSTON GEAR
700 SERIES
BOSTON GEAR

FIVE ON THE WALL

1998

details

left

SOAR EYES

1994

right

HUGE COUNTER

1996

Aluminum, acrylic,

electronics, LEDs

72 x 48 x 2 inches

far right

FIVE ON THE WALL

1998

detail

ARCHIMEDES, ET AL.
(Five historical
computations of π)
1996
Computers, software,
electronics, acrylic,
aluminum, LEDs
Five units:
66 x 108 x 2 inches
overall

THOUSAND, MILLION,
BILLION
1996
Computers, software,
aluminum, acrylic,
electronics, LEDs
Four units:
116 x 96 x 2 inches
overall

Instrumental or mechanical science is of all the noblest and the most useful, seeing that by means of this all animated bodies that have movement perform all their actions; and these movements are based on the center of gravity which is placed in the middle, dividing unequal weights, and it has dearth and wealth of muscles and also lever and counter lever. LEONARDO DA VINCI[1]

PLUG-IN PERSONALITIES

The art of robotics may have originated in the 15th century when Leonardo da Vinci began to explore the relationship between the organic and the mechanical. His aestheticized machinery set the stage for modern-day robotics. Until recently, machines were built to mimic the movements and behavior of humans and animals. The earliest robots were conceived of as machines to perform simple tasks that would allow humanity more time for intellectual pursuits and physical pleasures. Da Vinci's contemporary, Heinrich Cornelius Agrippa, wrote of self-propelled works of art in the form of the legendary automatons of Daedalus in Greek mythology, which carried out the tasks of cupbearers and servers at a banquet.[2]

The actual term "robot," used to designate a semi-autonomous mechanical creature, originated in a theatrical production. Robots were featured in the 1922 play *R.U.R.* (*Rossum's Universal Robots*) by Czechoslovakian author Karel Capek. Capek based the name for his mechanical anthropoids on *robota*, the Czech word for "forced labor," and on *rabota*, the Slavish word for "menial labor." In the play, the robots are a mechanical race built to serve humankind. Thus robots were born into slavery. Not surprisingly, the robots in the play eventually rebel against their masters and destroy them.

LIGHT, LANGUAGE, LUNAR LANDING, LAWS

Approximately 60 years later, Alan Rath began to make robot-like objects that serve no purpose at all. In his experiments, he stumbled upon an untrammeled path leading to electronic art. He found a way to overturn a conventional modernist attitude toward electronic sculpture as a means to service a mechanistic view of the world, with the ultimate aim to "control."[3] His work differs in many respects from most electronic art, and is closely allied with the tradition of kinetic sculpture stemming from Marcel Duchamp's 1913 *Bicycle Wheel*. The works of Alexander Calder, which Duchamp dubbed "mobiles" after a visit to the sculptor's studio; Jean Tinguely's exploding machines; and Rebecca Horn's mechanized contraptions are part of this tradition.

Mobility in the work of many kinetic artists depends on chance elements, such as air movement and temperature. In contrast, Rath programs his machines to "understand" a rudimentary language that enables them to respond directly to their environments without the artist's presence. For that reason his works are always theatrical; each piece has something of a personality — sometimes comic, at other times uncanny.

Born in Cincinnati in 1959, Rath learned the languages of art and science early in life. A childhood visit with his artist-mother to a Calder exhibition made an indelible impression. The 1969 moon landing was another pivotal childhood recollection for one who aspired to a career as a scientist. Since then, he has regarded the lunar landing module as one of the most beautiful works of sculpture produced in the 20th century. Rath is interested in experimental machinery that has not yet been "corrupted" by decoration. He has said that he prefers machines whose forms are tied to function—an efficient device still in an unrefined state, "before stylists get hold of the design." While a teenager, Rath designed lighting for a large-scale rock concert near his hometown. For him, electric rock guitar, blaring out of enormous speakers with screeching and grinding feedback noise, epitomized the analog sound of the '60s and '70s. Later, he made a number of speaker sculptures in homage to Jimi Hendrix. In general, machine sounds such as whirring motors, electronic buzzing and mechanical clicking are prominent features of Rath's work.

A passion for science and math led him to the Massachusetts Institute of Technology (MIT) some years later, where he earned a B.S. in electrical engineering. Sustaining a commitment to art, he spent long hours at the Center for Advanced Visual Studies, the Visual Language Workshop, and the Architecture Machine Group. He chose as his mentor the eminent electronics sculptor Otto Piene, an internationally renowned artist from Germany. Piene was acclaimed in the 1960s for his light sculptures, an art form that Rath adopted for his thesis project. Aside from Piene, other MIT instructors who made an impression on the young sculptor include the environmental theorist Leo Marx, and the psychologist Kenneth Kenniston, whose course "Technological Society and the Recovery of the Natural" Rath cites as a particularly rewarding experience.

Soon after graduation, Rath moved to California where he worked as an assistant to the Oakland-based artist Milton Komisar, known for his computerized light sculptures. Setting up a studio/garage/workshop in Oakland, Rath found himself content to settle in the Bay area, not far from Silicon Valley, where every imaginable piece of electronic equipment was readily available for purchase or loan. Rath held his first solo show of electronic sculptures at the Works Gallery in San Jose in 1986, and since then, his art and the laws that govern it have undergone a fascinating evolution.

CONTINUED ON PAGE 38

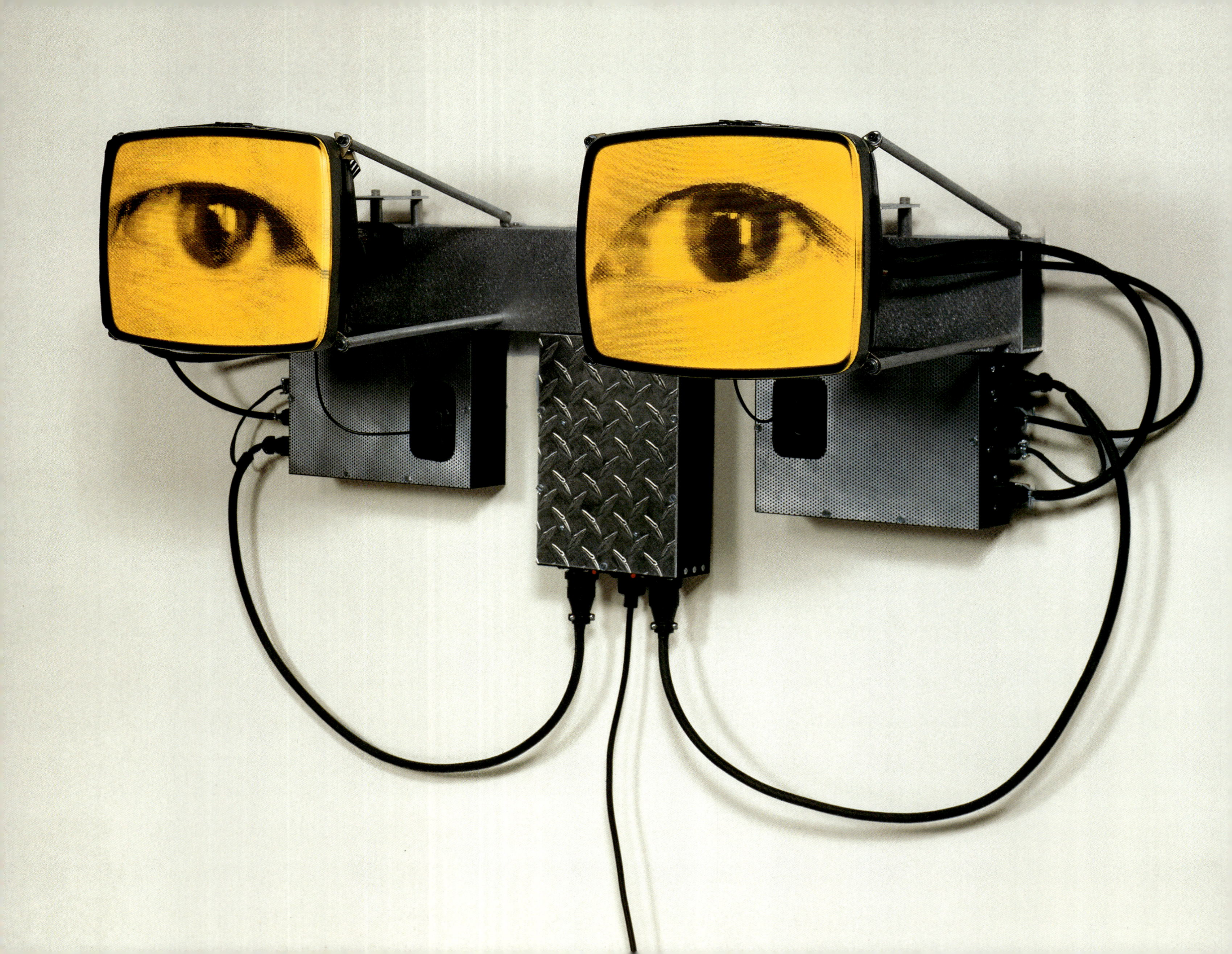

WATCHER

1998

Aluminum, electronics,

cathode ray tubes

24 x 42 x 13 inches

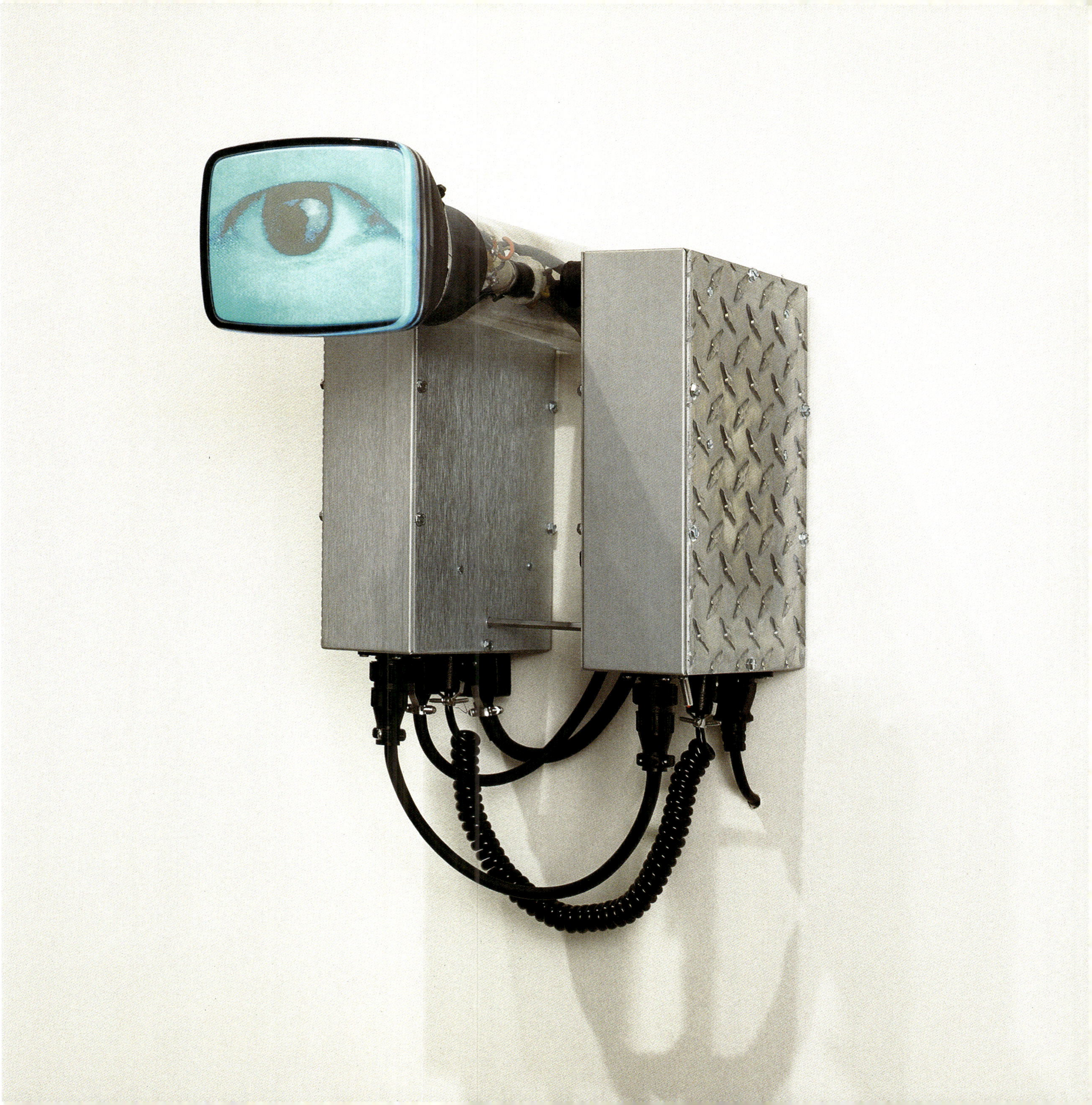

WALL EYE

1997

Computer, software,

aluminum, electronics,

cathode ray tube

14 x 9 x 13 inches

TENDER AND EROTIC MACHINES: THE FOUR GROUPS

The specialist in comprehensive design is an emerging synthesis of artist, inventor, mechanic, objective economist and evolutionary strategist. R. BUCKMINSTER FULLER [4]

Rath's interdisciplinary art calls to mind Buckminster Fuller's notion of a "comprehensive designer" whose integration of science and art could improve the plight of humankind. Similarly, Rath's work involves studies of human behavior, sociology, physics, chemistry, art history, and artificial intelligence. Such a broad range of interests would seem to manifest itself in varied art forms, and Rath himself recognizes in his output of the past decade four distinct bodies of work.

THE MOVING IMAGE. The first group includes sculptures featuring cathode ray tube (CRT) monitors, which show digitized video images of body movements, close-up views of blinking eyes, tongues protruding from mouths, tapping fingers, and hand signals. They recall highly stylized portraits or figure studies, albeit fragmented ones. Rath, however, has said that he is not as interested in anthropomorphic form as he is in behavior. "Washing machines don't look like somebody washing clothes," he has remarked.[5]

It must also be noted that Rath's CRT works are more closely allied with computer art than with video art. Some pieces share the playful qualities of certain works of video sculpture, such as those by Nam June Paik, an artist Rath admires. However, his objects diverge from video art in a number of ways. These self-contained systems using digital imagery differ significantly from Paik's installations and compositions that use transmitted analog video. In Rath's sculptures, the sequence of images on the monitors results from software programs that he has written especially for the individual piece. Thus, each element is thoroughly integrated into the whole, including the digitized imagery, which is always subordinate to the work's overall concept and design. On the other hand, some aspects of Rath's art may have spawned certain works by artists such as Tony Oursler, whose video-projection pieces rely on a kind of punning illusionism. Each Rath sculpture is part of an exploration of language and notions of communication. It is also important for Rath to be directly involved in every aspect of the work, not only its design, but also its construction. He starts with schematic drawings and designs all of the circuit boards and software.

Watcher, featuring a wall-mounted monitor showing a pair of eyes, and *Soar Eyes*, a dramatic airborne piece (suspended from the ceiling by wires), could be illustrative of Foucault's dictum that "the gaze that sees is the gaze that dominates."[6] The cold metallic armature of the pieces and the shifting glances of the enormous eyes corroborate the observation regarding the development of the notion of the gaze that "was triumphantly technological, whether in the architecture of Jeremy Bentham's *Panopticon* or in the lens of a camera. The disembodied authority extended power into sublime invisibility."[7] Imbuing the machine with lifelike attributes and human expression, Rath's sculptures are both comic and chilling. In an era of constantly expanding surveillance techniques, the work is a reminder that more often than not, "Big Brother *is* Watching You."

The spare and sleek sculpture *Triple Tongue Tree* is an approximately seven-foot-tall freestanding piece that resembles a coat rack. Three monitors and shelves bearing aluminum boxes that contain circuit boards are vertically arranged on a black metal tripod. With just a hint of Pop vulgarity, each monitor shows a video loop of a close-up of a mouth from which a tongue sticks out and wags around in a circle. Two screens on the top and bottom display a monochrome yellow mouth, while the middle screen shows a green mouth. It seems that Rath presents the image of a tongue as the locus of language. In its quirky semiology, the piece conveys an erotic charge and a certain irony that calls to mind Pop art images, such as Warhol's paintings of Marilyn Monroe's lips. Addressing a similar theme, the piece *Quartet* suggests an experiment in sign language. Here, a horizontal row of four wall-mounted monitors shows on each screen a yellow-tinted image of a hand waving or turning from side-to-side in varying patterns of stillness and motion. As the title indicates, the composition features a gracefully choreographed and coordinated ensemble.

Waiting III is a time piece of sorts. Placed on the floor, the work consists of a monitor connected by a metal tube to an aluminum carrying case. The screen displays a green-tinted image of fingers tapping endlessly on a tabletop, while a clock inset into the metal box shows only a red second hand whirling endlessly around and around. The humor of Rath's equation between the human and mechanical hand is soon dissipated by the work's relentless repetitions. Ultimately, the work conjures the myth of Sisyphus, a fascinating yet nightmarish view of eternity.

PARAGONS WITH A PULSE. The works in the second group are true kinetic sculptures, since they involve actual rather than virtual movement. Each work features large black speakers that emit inaudible low-frequency sounds, which cause the circular diaphragms to pulsate like the rhythmic breathing of animals. The sense of movement is subtle and sometimes almost imperceptible, which recalls the surreptitious motion of the elements in a sculpture by Pol Bury. Rath's *Pipes*, for example, is an elegant work in which large black speakers trimmed in orange are attached to thick black pipes that seem to have grown out of the ground. The work would suggest a cluster of giant mushrooms were it not for the gently throbbing speakers,

CONTINUED ON PAGE 42

QUARTET

1998

Aluminum, electronics,

cathode ray tubes

14 x 56 x 12 inches

which more closely resemble a batch of living coral. The piece also suggests a healthy specimen of some recently discovered species, a living, breathing paragon—a new life force, which is a half-plant, half-animal hybrid.

Quietly throbbing in midair, the three speakers in *Off the Wall III* are attached to long thin wires and suspended several feet from the ceiling. The wires undulate in graceful waves from the wall toward the center of the room, ending in the speakers' tender, quivering black membranes. The sculpture dramatically encroaches upon an area that viewers usually ignore: the space above their heads. While the work conveys the purposefulness of an elaborate PA system, its silently thumping speakers seem to have only subliminal messages to relay.

NUMBER SYSTEMS/GAMES. "Counting pieces" make up the third group. These flat rectangular, black panels are filled with glowing red LED numerals. They refer to our culture's obsession with numbers. Rath has etched into the circuit board of each piece a 1934 quotation by the writer Lewis Mumford, "In time keeping, in trading, in fighting, men counted numbers, and finally, as the habit grew, only numbers counted."

Strangely, the artist's only series of two-dimensional works are concerned with the fourth dimension. These statistical instruments pulsate to computer programs engaged in mathematical equations and puzzles. In a sense, they calculate the passage of time in the realm of time itself and according to time's own rules. In conceptual terms, it seems that Rath arrests time as he illuminates its passing. On a formal level, the works are akin to painting. The compositions relate to Constructivist art, for instance, and even to the blackboard drawings that the German artist Joseph Beuys often used in his performance/lectures.

In Rath's *Thousand, Million, Billion*, rows of large red numerals flashing against the dark grounds count up at the rate of one per second. It will take 31 years to reach one billion. *Huge Counter* is programmed to count in various ways, such as only in even or odd numbers, and in the prime numbers, at varying rates. Other works, such as the five-panel *Archimedes, et al. (Five historical computations of π)* span more than 2,000 years of calculations of pi. All of these hypnotic pieces convey a sense of the frantic pace of time. They imply a profound complexity yet they also harbor a jokey sense of humor that recalls certain works by Tatsuo Miyajima, the Japanese artist who attaches LED devices to hundreds of tiny toy cars that dart and bump about in a darkened room. A counting piece by Rath could also be seen as a heuristic or oracular device, one that promises to reveal the answer to cosmic questions or maybe just a winning number at a Bingo game.

Of all of Rath's works, the counting pieces are perhaps the most directly related to his study of verbal and theoretical language. They constitute an homage to the mathematical language of the algorithm. Jeremy Campbell in his *Grammatical Man* defines an algorithm as "a special method of manipulating symbols, especially one that uses a single basic procedure over and over again. It converts certain quantities into other quantities using a finite number of transformation rules."[8] The word "algorithm" comes from Al-Khowarizmi, the ninth-century mathematician and astronomer who invented algebra, and who was a member of the House of Wisdom in Baghdad, an academy of scientific thought.

In spite of their ultra high-tech qualities, Rath's counting pieces conjure certain romantic notions of an ancient astronomer mapping the stars in search of truth. Like emblematic shields in a digital battle with eternity, the pieces flicker in steadfast opposition to the tyranny of time.

HUMAN MORTALITY/ROBOT MORALITY. The fourth group contains the robotic sculptures. In these haunting pieces, Rath makes his most persuasive bid to humanize technology. His machines are not aggressive monsters, nor are they passive or subservient beasts. While their movements hint at human behavior and social interactions, they are not anthropomorphic. Elements in these complex pieces dance together in sensuous rapture or duel without touching in intense mock battles, but they always act in accord with their own cyber-sensibilities.

In numerous pieces, the various components resemble abstracted limbs. Long strips of aluminum, like slender bands in certain George Rickey sculptures, appear as motorized metal "arms" in Rath's work. They gesture wildly in *Robot Dance* and *Five on the Wall*. Other works, such as *Inchworm* and *One Track Minds* glide across the floor in graceful strides or move in spastic thrusts. *Inchworm* calls to mind one of Walter de Maria's cool stainless-steel sculptures. Rath's sleek, minimalist composition is made of two six-foot metal beams hinged together to form an upside-down "V." The object inches along the floor at the nonchalant pace of a giant garden slug.

Rath's robots seem to exercise cautious optimism as they venture out into the world. In a sense they are akin to the early conceptions of robots as good-deed doers that appeared in Japan soon after World War II. In the early 1950s, Osamu Tezuka created "Mighty Atom" (or "Astro Boy" as he appeared in the West) to be a nearly perfect robot who strove to become more human, and also to be an interface between two different cultures—that of man and that of machine.[9] While Rath's machines may be high-tech, it is important to note that they have little in common with sci-fi robots. They stand apart from the pugnacious monsters

CONTINUED ON PAGE 46

WAITING III

1998

Aluminum, electronics,

clock, cathode ray tube

18 x 42 x 30 inches

TRIPLE TONGUE TREE

1998

Tripod, aluminum,

steel, electronics,

cathode ray tubes

81 x 41 x 39 inches

made by sculptor Kenji Yanobi, or sci-fi cinema creatures.[10] Rath's sleek, silvery machines are neither futuristic nor nostalgic. Instead, the artist aims for a kind of timeless classicism; he strives to create a machine of perfect balance, proportion, and line.

A reflection of humanity's epic struggle against time and mortality, classical art spans millennia. Machines usually don't resemble the works by Phidias that adorn the Parthenon, but an intelligent and well-built robot may prove to be just as immortal. It seems fitting that Rath's first robotic sculptures should be wall-reliefs, similar in format to many ancient sculptures, such as the carved pylons of Egyptian temples, the Ara Pacis in Rome, or Mayan stelae found throughout Central America. *Robot Dance* (1995), Rath's first venture into robotics, is a spirited wall piece consisting of two approximately five-foot-long motorized metal arms hanging adjacent to one another about five feet from the floor. The arms come in three sections — two long and one short — that reference upper arms, forearms and fingers. Connected by elbow and wrist joints, the sections swivel up and down and rotate. Stretched to their limit, the arms nearly touch like the fingers of God and Adam in Michelangelo's painting on the Sistine Chapel ceiling.

A pair of wheeled elements with flailing arms in *One Track Minds* lunge at each other with jabbing arm movements as if they were opponents in a fencing dual or lovers in a passionate mating ritual. Responding to built-in motion sensors, the computerized vehicles perform with the jerky and quirky movements of Javanese shadow puppets. Similarly theatrical are the synchronized movements in *Five on the Wall*, where the high-kicking, gyrating elements have been likened to the precision line dances of the Rockettes.[11] In *California Landscape*, Rath sets up a scenic tableau. An aluminum wall element, which looks like a large eye or a stylized sun or moon, consists of a metal disk traversed by curving metal rods. A corresponding floor element, with two gyrating arms, slides back-and-forth across the room on tracks, like an abstracted car speeding along a California freeway.

Friends and Acquaintances is an intricate and wildly exuberant work in which Rath perhaps most convincingly conflates the human, organic, and mechanical. The five elements in the piece — three freestanding tripods and two wall-mounted metal boxes — interact with each other in a way that hints at sexual activity and verbal communication. Revealing brightly lit, warm-red interiors, the boxes open and close in response to other elements in the sculpture. Long, rolled-up metal tongues unfurl and protrude in a comic, though rather lurid way to penetrate the open boxes attached to the wall. No lubricating fluids pass from one machine to another, so the implied sexual activity is apparently "safe." Yet, an issue of robot morality is suddenly called into question.

In *Rover*, the artist's most recent robotic sculpture, motion sensors allow interaction with visitors. The squat, wheeled vehicle emerges from beneath a canopied recharging station as guests arrive. An eye on a small CRT screen at the end of a long rod inspects the intruder like a suspicious watchdog. But there is no sense of menace to this creature. In this piece and in all of his other works, Rath conveys a technological world that is inviting rather than fearful. He has gone to great lengths to counter the technological alienation described by Leo Marx, who wrote "To satisfy the imperatives of technological society, men are called upon to endure an intolerable curbing of their spontaneous, erotic and personal selves." [12]

Neither utopian nor dystopian, Rath's visionary art encompasses technology as a cohesive language, one whose subtlety and expressive power matches that of any other art medium. Alan Rath has already made an enormous contribution to machine culture, and today he stands among the most important artists whose works redefine what art is and what it could be.

DAVID EBONY

1
The Notebooks of Leonardo da Vinci, Pamela Taylor, ed., New York: Plume/New American Library, 1971, 197-98.

2
The Lure of Antiquity and the Cult of the Machine, Horst Bredekamp, Princeton: Marcus Wiener, 1995, 48-49.

3
Passages in Modern Sculpture, Rosalind E. Krauss, Cambridge: MIT, 1981, 220.

4
Ideas and Integrities, R. Buckminster Fuller, New York: Collier, 1963, 176.

5
Rath interviewed by Antonio López, *Pasatiempo, The New Mexican*, Santa Fe, NM, Oct. 30, 1998.

6
Electronic Culture: Technology and Visual Representation, Timothy Druckrey, New York: Aperture, 1997, 18.

7
Ibid., 18-19.

8
Grammatical Man: Information, Entropy, Language and Life, Jeremy Campbell, New York: Simon & Schuster, 1982, 130.

9
Inside the Robot Kingdom: Japan Mechatronics and the Coming Robotopia, Frederik L. Schodt, New York: Kodansha, 1990, 75.

10
Winona Ryder playing a robot in the 1998 film *Alien Resurrection*, for example.

11
Alan Rath: New Robotic Sculpture, exhibition flyer essay by Renny Pritikin, San Francisco: Yerba Buena Center for the Arts, 1998.

12
The Machine in the Garden: Technology and the Pastoral Ideal in America, Leo Marx, New York: Oxford, 1964, 178.

DAVID EBONY is an editor of *Art in America*. He is also a contributor to the on-line magazine *ArtNet*. He lives and works in New York City.

PIPES

1997

Plastic, aluminum, electronics, speakers

56 x 34 x 26 inches

OFF THE WALL III

1995

Aluminum, electronics, speakers

dimensions variable

up to 30 feet long

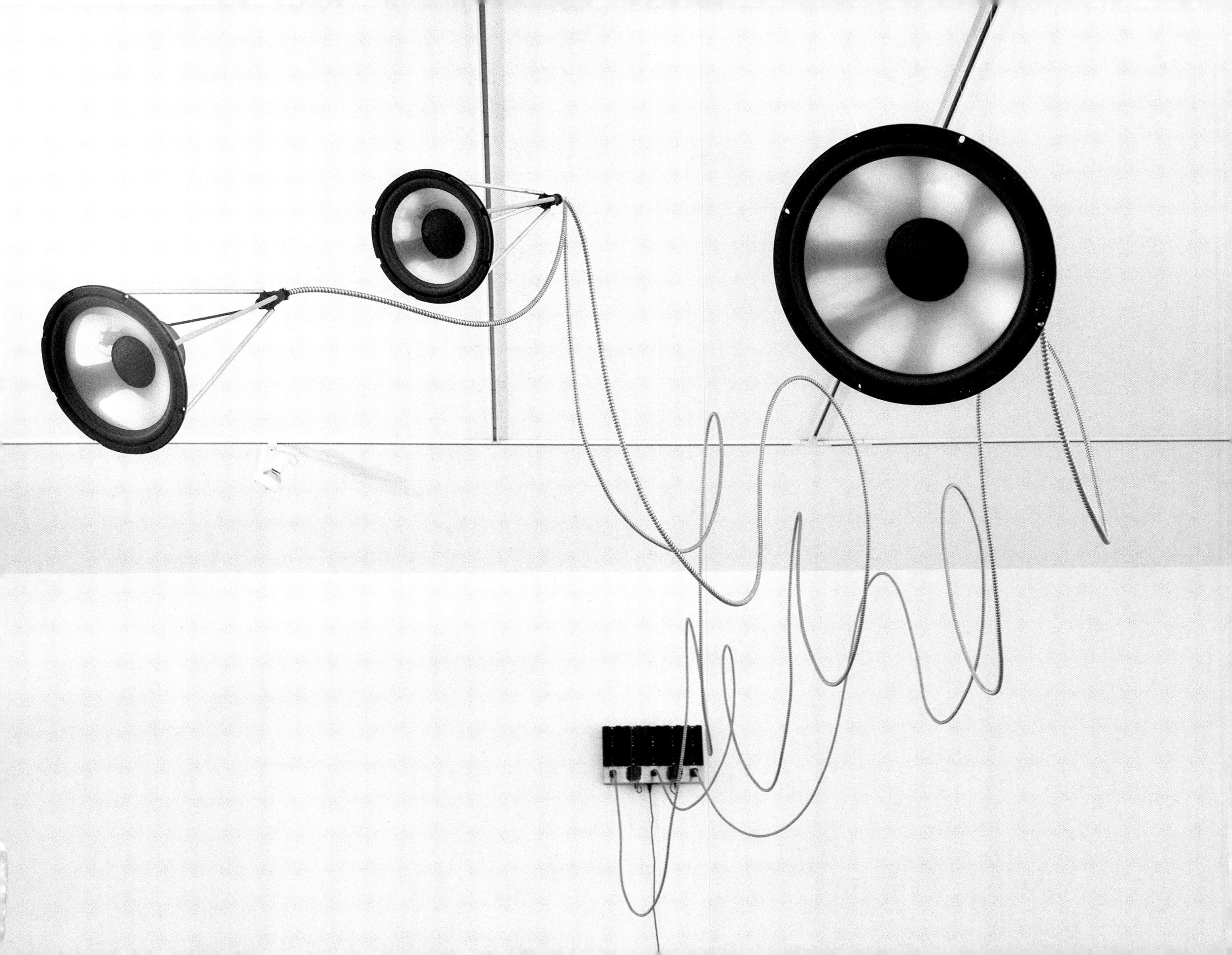

ONE TRACK MINDS

1998

Computers, software,

aluminum, motors

Track: 40 feet long

Carts: 11 x 4 x 2 feet,

13 x 3 x 2 feet

ONE TRACK MINDS

1998

installation views

sketch for

ONE TRACK MINDS

1997

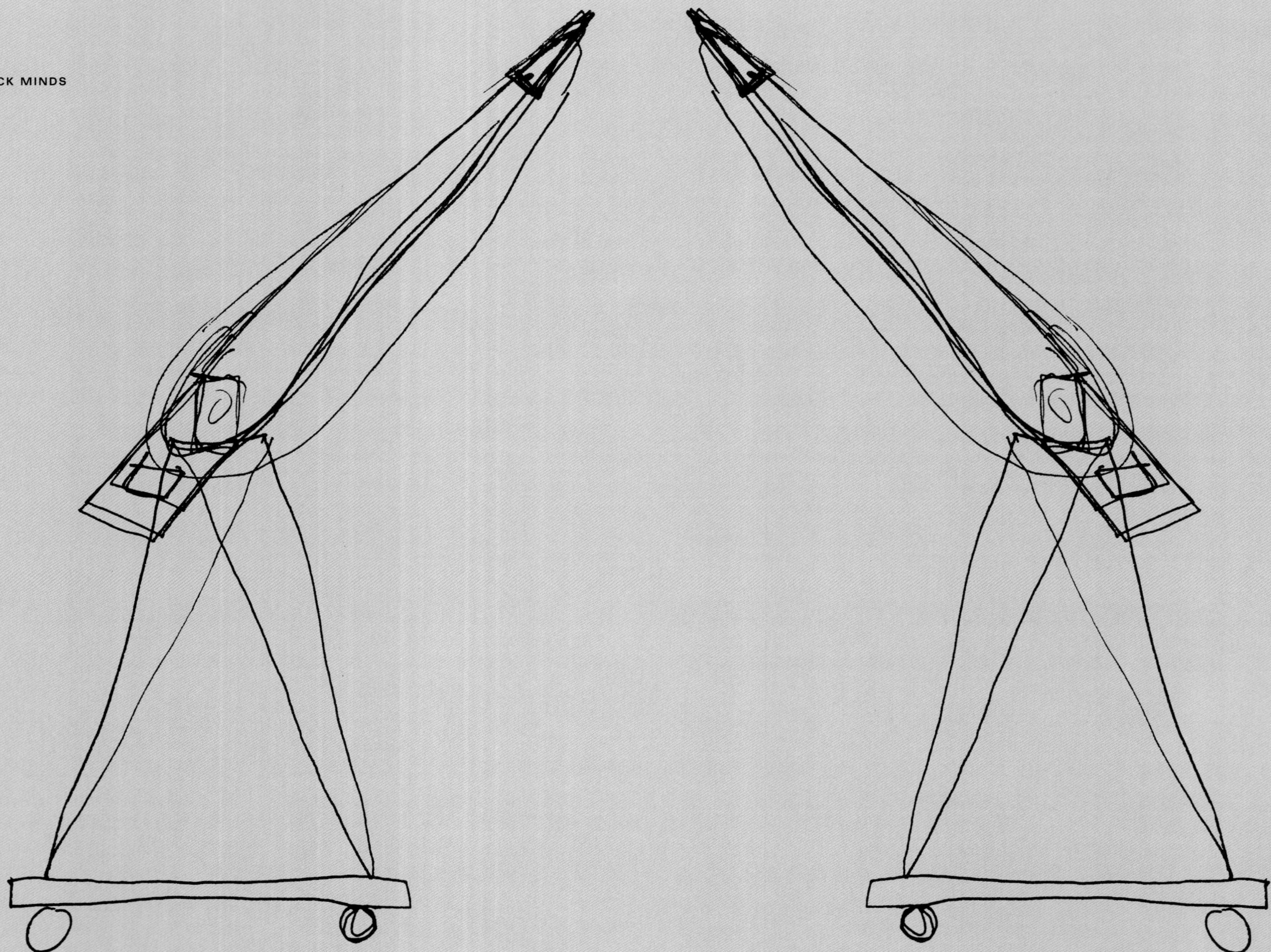

EDUCATION

1982 Massachusetts Institute of Technology, Cambridge, Massachusetts, Bachelor of Science in Electrical Engineering

SELECTED SOLO EXHIBITIONS

1998 SITE Santa Fe, Santa Fe, New Mexico (catalogue)
traveling in 1999 to: Austin Museum of Art, Austin, Texas; Scottsdale Museum of Contemporary Art, Scottsdale, Arizona
Yerba Buena Center for the Arts, San Francisco, California
The Huntington Beach Arts Center, Huntington Beach, California
Haines Gallery, San Francisco, California
Dorfman Projects, New York, New York

1996 The Aspen Art Museum, Aspen, Colorado
Nelson Gallery, University of California, Davis, California
Haines Gallery, San Francisco, California

1995 Contemporary Arts Museum, Houston, Texas
Haines Gallery, San Francisco, California
Palo Alto Cultural Center, Palo Alto, California

1994 Hiroshima City Museum of Contemporary Art, Hiroshima, Japan
Worcester Art Museum, Worcester, Massachusetts
John Weber Gallery, New York, New York
Carl Solway Gallery, Cincinnati, Ohio

1993 Gallery of Art, Johnson County Community College, Overland Park, Kansas
Southeastern Center for Contemporary Art, Winston-Salem, North Carolina

1992 Galerie Hans Mayer, Düsseldorf, Germany
Dorothy Goldeen Gallery, Santa Monica, California
University of Missouri, St. Louis, Missouri (catalogue)

1991 Walker Art Center, Minneapolis, Minnesota
traveled in 1992 to: Museum of Contemporary Art, Chicago, Illinois; Institute of Contemporary Art, Philadelphia, Pennsylvania; Center for the Fine Art, Miami, Florida; and The Contemporary Museum, Honolulu, Hawaii
Carl Solway Gallery, Cincinnati, Ohio

1990 San Jose Museum of Art, San Jose, California
Dorothy Goldeen Gallery, Santa Monica, California
Gallery Paule Anglim, San Francisco, California

SELECTED SOLO EXHIBITIONS (CONTINUED)

1989 Katia Lacoste Gallery, San Jose, California

1988 New Langton Arts, San Francisco, California

1987 ProArts, Oakland, California
Memorial Union Art Gallery, University of California, Davis, California

1986 Artists' Television Access, San Francisco, California
WORKS / San Jose, San Jose, California

1984 Center for Advanced Visual Studies, MIT, Cambridge, Massachusetts

SELECTED GROUP EXHIBITIONS

1999 Weston Art Gallery, Cincinnati, Ohio, *Digital Dialogues*

1998 Kunsthalle Wien, Vienna, Austria, *Crossings: Art to See and Hear*
Wexner Center for the Arts, Columbus, Ohio, *Body Mécanique* (catalogue)
Haines Gallery, San Francisco, California, *Plugged In*
The Transamerica Pyramid Building, San Francisco, California, *Perspectives on Infinity*

1997 San Jose Museum of Art, San Jose, California, *Alternating Currents: American Art in the Age of Technology*
Contemporary Arts Museum, Houston, Texas, *Finders / Keepers* (catalogue)
Art Academy of Cincinnati, Cincinnati, Ohio, *Digital Decisions*
The Transamerica Pyramid Building, San Francisco, California, *About Face*

1996 La Lonja, Palma de Mallorca, Spain, *Art Medial* (catalogue)
Otso Gallery, Espoo, Finland, *Metamachines—Where is the Body*
Delaware Center for the Contemporary Arts, Wilmington, Delaware, *A Chaos of Delight*
Fotouhi Cramer Gallery, New York, New York, *The Box Exhibition*
Haines Gallery, San Francisco, California, *Matters Of The Heart*
The Aldrich Museum of Contemporary Art, Ridgefield, Connecticut, *Landscape Reclaimed*
Dorothy Goldeen Gallery, Santa Monica, California, *A Selection of Classics*

1995 San Francisco Museum of Modern Art, San Francisco, California, *Recent Acquisitions*
Maryland Institute, College of Art, Baltimore, Maryland, *Project: Exhibit*
The Huntington Beach Art Center, Huntington Beach, California, *Veered Science* (catalogue)
Richmond Art Center, Richmond, California, *Framing Exposure*
The Aldrich Museum of Contemporary Art, Ridgefield, Connecticut, *Inside Out* (catalogue)
New Jersey Center for Visual Arts, Summit, New Jersey, *The Artful Message: Contemporary Video* (catalogue)
University of Massachusetts, Amherst, Massachusetts, *In Vivo*
Capp St. Project, San Francisco, California, *Old Glory* (catalogue)

Art Museum of South Texas, Corpus Christi, Texas, *Between Reality and Abstraction: California Art at the End of the Century* (catalogue)
Track 16 Gallery, Santa Monica, California, *L.A. International Biennial Invitational*
Newport Harbor Art Museum, Newport Beach, California, *Machine*
The Contemporary Arts Center, Cincinnati, Ohio, *It's Only Rock and Roll: Rock and Roll Currents in Contemporary Art* (catalogue)
Gallery of Art, Johnson County Community College, Overland Park, Kansas, *View / Review*
Art Miami, Miami, Florida, *Art on the Eve of the Millennium*

1994 Spiral Art Center, Tokyo, Japan, *Of the Human Condition: Hope and Despair at the End of the Century* (catalogue)
The MIT Museum, Cambridge, Massachusetts, *25 Years of the Center for Advanced Visual Studies*
San Antonio Museum of Art, San Antonio, Texas, *Synesthesia: Sound and Vision in Contemporary Art* (catalogue)
Haines Gallery, San Francisco, California, *Issues of Image*
Table, Lamp, & Chair, Portland, Oregon
The Contemporary Arts Center, Cincinnati, Ohio

1993 Akademie der Künste, Berlin, Germany, *Akademie '93* (catalogue)
International Center of Photography, New York, New York, *Iterations* (catalogue)
Museum of Art, Palm Beach Community College, Palm Beach, Florida, *Things That Go Bump in the Night*
Southern Exposure, San Francisco, California, *Science Fair*
John Weber Gallery, New York, New York, *Toward A New Vision*
Haines Gallery, San Francisco, California, *Body Parts*
The Contemporary Arts Center, Cincinnati, Ohio, *The Figure as Fiction* (catalogue)
The Wood Street Galleries, Pittsburgh, Pennsylvania, *Art in the Age of Information*

1992 Padiglione d'Arte Contemporanea, Ferrara, Italy, *Polise* (catalogue)
International Center of Photography, New York, New York, *Special Collections* (catalogue)
Alvar Aalto Museum, Jyväskylä, Finland, *Lumo '92: Space in Picture, Picture in Space* (catalogue)
Thread Waxing Space, New York, New York, *The Interactive Show*
The School of the Museum of Fine Arts, Boston, Massachusetts, *Alan Rath, Larry Johnson, Laurie Simmons*
The Aspen Art Museum, Aspen, Colorado

1991 Whitney Museum of American Art, New York, New York, *1991 Biennial Exhibition* (catalogue)
Denver Art Museum, Denver, Colorado, *Visions / Revisions*
Amy Lipton Gallery, New York, New York, *Breathing Room*
The Contemporary Arts Center, Cincinnati, Ohio, *Mechanika* (catalogue)
Milwaukee Art Museum, Milwaukee, Wisconsin
Germans van Eck Gallery, New York, New York

SELECTED GROUP EXHIBITIONS (CONTINUED)

1990 San Francisco Museum of Modern Art, San Francisco, California, *Bay Area Media*
Stadtmuseum Siegburg, Siegburg, Germany, *Lichtkunst und Klangobjekte*
Cité des Arts et des Nouvelles Technologies, Montreal, Canada, *Image du Futur, '90* (catalogue)
Oakland Museum, Oakland, California, *Oakland Artists, '90*
University of California at Santa Barbara, Santa Barbara, California, *P.U.L.S.E. 2* (catalogue)
Miami University Art Museum, Oxford, Ohio, *Wired: Art That Moves*
San Diego Museum of Art, San Diego, California, *Contemporary Collectors*

1989 Los Angeles Contemporary Exhibitions, Los Angeles, California, *Self / Evidence* (catalogue)
Ars Electronica, Linz, Austria, *Im Netz der Systeme* (catalogue)
Visiona, Zurich, Switzerland (catalogue)
Terrain Gallery, San Francisco, California, *Matter, Anti-matter: Defects in the Model*
Carl Solway Gallery, Cincinnati, Ohio, *Sculpture*
Intermedia Arts, Minneapolis, Minnesota, *The Technological Imagination: Machine in the Garden of Art* (catalogue)
Artspace, San Francisco, California, *1989 Sculpture Awards*
Dorothy Goldeen Gallery, Santa Monica, California, *Fresh Visions II—The Photographic Image in 2-D and 3-D*
Sesnon Gallery, University of California at Santa Cruz, Santa Cruz, California, *Alternative Applications: Computer Technology in the Arts*
The Contemporary Arts Center, Cincinnati, Ohio
Security Pacific Gallery, Costa Mesa, California, *Media Talk* (catalogue)
Museum Folkwang, Essen, West Germany, *Gefangene Bilder, Fluchtiges Gedachtnis, Digital Fotografie*
Syntex Gallery, Palo Alto, California, *Left Brain / Right Brain*

1988 San Francisco Museum of Modern Art, San Francisco, California, *SECA Video Invitational*
Badischer Kunstverein, Karlsruhe, West Germany, *Otto Piene und das CAVS* (catalogue)
Siggraph '88, Atlanta, Georgia, *Interactive Techniques*
University of Wisconsin, Eau-Claire, Wisconsin, *New Art*
San Francisco State University, San Francisco, California, *Countervisions: Pioneers in Bay Area Art*
San Francisco Camerawork Gallery, San Francisco, California, *Digital Photography* (catalogue)

1987 Photographic Resource Center, Boston, Massachusetts, *High Tech, New Pop*
Everson Museum of Art, Syracuse, New York, *Computers and Art* (catalogue)
California Polytechnic State University, San Luis Obispo, California, *Summer Arts Celebration*
1160 Bryant St., San Francisco, California, *X: The Unknown*
Santa Rosa Community College Art Gallery, Santa Rosa, California, *Sculpture: Four Views*
Walnut Creek Civic Arts Gallery, Walnut Creek, California, *Lights On!*
WORKS / San Jose, San Jose, California, *Eye Magic*

1986 San Francisco International Video Festival, San Francisco, California, *Four on the Floor*
Los Angeles County Fair, Pomona, California, *New Visions*
Berkeley Art Center, Berkeley, California, *Art and the Computer*

1985 Sausalito Arts Festival, Sausalito, California

1984 Boston "First Night" Celebration of the Arts, Boston, Massachusetts

FELLOWSHIPS

1994 John Simon Guggenheim Memorial Foundation, New York, New York

1988 National Endowment for the Arts, Washington, DC

PUBLIC COLLECTIONS

Cincinnati Art Museum, Cincinnati, Ohio

Denver Art Museum, Denver, Colorado

Hara Museum of Contemporary Art, Tokyo, Japan

Museum für Holographie & neue visuelle Medien, Cologne, Germany

Oakland Museum, Oakland, California

Progressive Corporation, Mayfield Heights, Ohio

San Diego Museum of Contemporary Art, La Jolla, California

San Francisco Museum of Modern Art, San Francisco, California

Santa Barbara Museum of Art, Santa Barbara, California

Walker Art Center, Minneapolis, Minnesota

Whitney Museum of American Art, New York, New York

Worcester Art Museum, Worcester, Massachusetts

CHECKLIST

CALIFORNIA LANDSCAPE
1998
Computers, software, electronics, aluminum, motors
Two units:
12 x 25 x 12 feet overall
Courtesy of the artist and Haines Gallery, San Francisco.
Commissioned by Yerba Buena Center for the Arts, San Francisco

FIVE ON THE WALL
1998
Computers, software, electronics, aluminum, motors
Five units:
16 x 55 x 2 feet overall
Courtesy of the artist and Haines Gallery, San Francisco.
Commissioned by Yerba Buena Center for the Arts, San Francisco

FRIENDS AND ACQUAINTANCES
1998
Computers, software, electronics, aluminum, motors
Five units:
12 x 33 x 14 feet overall
Courtesy of the artist and Haines Gallery, San Francisco.
Commissioned by Yerba Buena Center for the Arts, San Francisco

INCHWORM
1998
Computers, software, electronics, aluminum, motors
Dimensions variable, approx. 73 inches high
Courtesy of the artist and Haines Gallery, San Francisco.
Commissioned by SITE Santa Fe

ONE TRACK MINDS
1998
Computers, software, aluminum, motors
Track: 40 feet long
Two units: 11 x 4 x 2 feet, 13 x 3 x 2 feet
Courtesy of the artist and Haines Gallery, San Francisco

QUARTET
1998
Aluminum, electronics, cathode ray tubes
14 x 56 x 12 inches
Courtesy of the artist and Haines Gallery, San Francisco

ROVER
1998
Computers, software, electronics, aluminum, motors
approx. 44 x 38 x 41 inches
Courtesy of the artist and Haines Gallery, San Francisco
Commissioned by SITE Santa Fe

TRIPLE TONGUE TREE
1998
Tripod, aluminum, steel, electronics, cathode ray tubes
81 x 41 x 39 inches
Courtesy of the artist and Haines Gallery, San Francisco

WAITING III
1998
Aluminum, electronics, clock, cathode ray tube
18 x 42 x 30 inches
Courtesy of the artist and Haines Gallery, San Francisco

WATCHER
1998
Aluminum, electronics, cathode ray tubes
24 x 42 x 13 inches
Courtesy of the artist and Haines Gallery, San Francisco

PIPES
1997
Plastic aluminum, electronics, speakers
56 x 34 x 26 inches
Courtesy of the artist and Haines Gallery, San Francisco

WALL EYE
1997
Computer, software, aluminum, electronics, cathode ray tube
14 x 9 x 13 inches
Courtesy of the artist and Haines Gallery, San Francisco

ARCHIMEDES, ET AL. (FIVE HISTORICAL COMPUTATIONS OF π)
1996
Computers, software, electronics, acrylic, aluminum, LEDs
Five units:
66 x 108 x 2 inches overall
Courtesy of the artist and Haines Gallery, San Francisco

HUGE COUNTER
1996
Aluminum, acrylic, electronics, LEDs
72 x 48 x 2 inches
Courtesy of the artist and Haines Gallery, San Francisco

THOUSAND, MILLION, BILLION
1996
Computers, software, aluminum, acrylic, electronics, LEDs
Four units:
116 x 92 x 2 inches overall
Courtesy of the artist and Haines Gallery, San Francisco

OFF THE WALL III
1995
Aluminum, electronics, speakers
Dimensions variable, up to 30 feet long
Courtesy of the artist and Haines Gallery, San Francisco

ROBOT DANCE
1995
Computer, software, electronics, aluminum, motors
Two units:
10 x 20 x 3 feet overall
Courtesy of the artist and Haines Gallery, San Francisco.
Commissioned by Yerba Buena Center for the Arts, San Francisco

SOAR EYES
1994
Aluminum, acrylic, electronics, cathode ray tubes
9 x 32 x 45 inches
Collection of Dr. Robert C. and Nancy Magoon

Many people helped make this exhibition possible. First I'd like to thank my long time studio assistant Bruce Cannon who has in some way touched almost every piece in this exhibition. Neil Grimmer, Nathaniel Taylor, and Paul Stout helped fabricate the new robotic pieces. The John Simon Guggenheim Foundation provided a fellowship that enabled my earliest explorations into computer-controlled mechanical sculpture. The Gerbode Foundation supported the construction of *Robot Dance*. My thanks to Renny Pritikin and the Yerba Buena Center for the Arts for supporting and premiering *Friends and Acquaintances*, *California Landscape*, *Five on the Wall*, and *One Track Minds*.

The entire staff of SITE Santa Fe has been extremely diligent and helpful in presenting this most demanding exhibition. I especially wish to thank Louis Grachos for facilitating this exhibition with its 40 motors and 51 computers, and for commissioning the new pieces *Rover* and *Inchworm*. Karen Christiansen graciously lent her own computer monitor to the show when I ended up one short. Craig Anderson and his crew expertly handled the installation of the works, and Melissa Dubbin has patiently monitored the daily performance of the robots.

Cheryl Haines of Haines Gallery has been critical in realizing this project, and I thank her for continuing to put up with my crazy ideas. A number of collectors have supported my work in ways that transcend collecting, among them are Kent and Vicki Logan. Similarly, Tom Patchett has assisted in many ways, as have David and Isabel Breskin. My thanks to Bob and Nancy Magoon for lending to the show on short notice. I am grateful to Toby Devan Lewis and Ray Graham for their support of this exhibition. It is a real pleasure to have Murray Gell-Mann and David Ebony make written contributions to this catalogue and to have Sarah King coordinate the catalogue production.

Finally, I wish to thank Schockley, Brattain, and Bardeen for inventing the transistor, without which, who knows what I would be doing now.

ALAN RATH

SITE SANTA FE would like to thank the following individuals and organizations for their generous support:

Toby Devan Lewis, Elizabeth Firestone Graham Foundation, The Max and Victoria Dreyfus Foundation Inc., The Shifting Foundation, Smart Art Press, and The Vicki and Kent Logan Family Foundation.

All work, unless otherwise noted, courtesy of the artist and Haines Gallery, San Francisco. *Robot Dance*, *Five on the Wall*, and *California Landscape* are courtesy of the artist, Haines Gallery, San Francisco, and Yerba Buena Center for the Arts, San Francisco. *Soar Eyes* is from the collection of Dr. Robert C. and Nancy Magoon, Aspen, Colorado.

ALSO FROM SMART ART PRESS:

What can we say about an artist who so deftly conflates the realms of technology and nature. Is it a man, a plant, a machine? Who cares when the life force is so strong and the resulting works of art so enchanting?

Alan Rath: Plants, Animals, People, Machines

For Rath, technology is both the subject and medium of his artistic inventions. With a substantial dose of humor, his sculptures—part botanical, part anthropomorphic, part machine—operate as emotional and psychological metaphors, suggesting that, like language, technology also shapes the way we perceive the world. With essays by Peter Boswell of the Walker Art Center; Louis Grachos of the Museum of Contemporary Art, San Diego; John Hanhardt, video and film curator at the Whitney Museum; and Donna Karkavy of the Worcester Art Museum.

1995, SOFTCOVER, 7 ½ X 10 IN. 48 PP. 18 COLOR, 37 B&W ILLUSTRATIONS, PLUS 3 PG COLOR TIP-IN. $20

Virtual Rath

Limited Edition screensaver by this virtuosic electronic sculptor with his humorous signature subjects: searching eyes, sniffing noses, and licking tongues. Signed and numbered in an edition of 750 packaged with *Alan Rath: Plants, Animals, People, Machines*.

1997, $100

Liza Lou

This jewel-like book documents *The Kitchen* and *Back Yard*, Lou's extravagant domestic environments made entirely of millions of tiny glass beads.

1997, SOFTCOVER, 6 ¾ X 6 ¾ IN. 64 PP. 35 COLOR REPRODUCTIONS. $20

Beattie & Davidson

For painters Drew Beattie and Daniel Davidson, "two heads are better than one, four hands better than two." Since 1989, the New York-based duo has pursued an unusual creative partnership, working together and in alternation on their paintings and collages. The artists' "exquisite corpse"-influenced aesthetic embraces accident and contradiction, making their paintings stylistic free-for-alls that spontaneously careen from the representational to the abstract to the cartoonlike. Essays by Larry Rinder and David Humphrey.

1998, SOFTCOVER, 11½ X 9½ IN. 72 PP. 50 COLOR AND 5 B&W REPRODUCTIONS. $20

sketch for

CALIFORNIA LANDSCAPE

1997

SITE SANTA FE

STAFF

Louis Grachos
Director & Curator

Craig Anderson
Exhibitions & Facilities Manager

Kimberly Botza
Director's Assistant

Jodi Carson
Assistant Director of Publications

Christina Cassidy
Public Relations Coordinator

Karen Christiansen
Assistant Director of Administration & Development

Martha DeFoe
Accounting & Administrative Assistant

Marianne Dell
Public Relations Assistant

Melissa Dubbin
Registrar

Dennis Esquivel
Facilities & Maintenance

Margaret Grant
Marketing & Membership Associate

Sarah King
Catalogue Editor & Special Projects

Graham Murtough
Marketing Assistant

Chris Nail
Education Coordinator

Judith Podmore
Docent Coordinator

Erin Shirreff
Publications Assistant

Siobhan Spain
Visitor Services Coordinator

Bryan Walker
Facilities & Events Assistant

Preparators:
Bruce Clotworthy
Steve Fowler
James Holmes
Terri Rolland
David Servoss
Peter Sprunt
Tom Tiegler
Colin Zaug

Docents:
Madelin Coit
Kathryn Davis
Maggie Furnas
Ann Hosfeld
Lisa Kirk
Nancy Kriebel
Nick Livaich
Stephanie Lynch
Donald Meyer
Sally Mittler
Chris Nail
Judith Podmore
Eleanor Rappe
Celia Rumsey
Ginna Sloane
Rusty Spicer
Signe and Joe Stuart
Nancy Sutor
Sarah Tyson
Peter Wiedmann
Judy Youens
Barbara Zusman

Interns:
Betsy Davis
Meaghan Kent
Vivi Valentine
Jennifer Wells

BOARD OF DIRECTORS

Joann Phillips
Chairman

Bobbie Foshay-Miller
President

Robert Denison
Treasurer

William Miller
Secretary

Richard Barrett
James F. Fitzpatrick
Emily Fisher Landau
Nancy Magoon
John L. Marion
Balene McCormick
Ronald A. Pizzuti
Dede Schuhmacher
Louisa Stude Sarofim

INTERNATIONAL ADVISORY COMMITTEE

Michael Auping
Bruce W. Ferguson
Michael Govan
Agnes Gund
John R. Lane
Lars Nittve
Gifford Phillips
Griselda Pollack
Ned Rifkin

NEW MEXICO ADVISORY COMMITTEE

Donald Meyer
Co-Chair

Larry Ogan
Co-Chair

Jonathon Abrams
Jan Ernst Adlmann
Stuart Ashman
Dottie Barrett
Ellen Bradbury-Reid
Bunny Conlon
Jill Z. Cooper
Guy Cross
Marjorie Devon
Linda Durham
Robert Ellis
Jim Enyeart
Zane Fischer
Megan Fox
Tom Guralnick
Dwight Hackett
Saskia Hamilton
Charlotte Jackson
James Kelly
George King
Forrest Moses
Patricia Garcia Salazar

Copublished by SITE Santa Fe
and Smart Art Press
on the occasion of the exhibition
ALAN RATH: ROBOTICS
October 31, 1998 – January 24, 1999

Organized by SITE Santa Fe
Curated by Louis Grachos

ISBN#0-9650583-5-2
Library of Congress Card Catalog
Number: 98-83235

Design: Joseph Guglietti Design
Editor: Sarah S. King
Copyeditor: Rebecca Friedman
Photographs are by Herbert Lotz
with the following exceptions:
Ben Blackwell: pp 4,16, 17, 36,
44, 45, 48
Susan Scott: p 24

front cover:
ROVER, 1998, detail
back cover:
ROVER, 1998, installation view

Traveling Exhibition dates:

AUSTIN MUSEUM OF ART
April 17 – June 13, 1999

SCOTTSDALE MUSEUM OF CONTEMPORAY ART
September 10, 1999 – January 2, 2000

SITE Santa Fe
1606 Paseo de Peralta
Santa Fe, New Mexico
87501

505-989-1199
FAX-998-1188

www.sitesantafe.org

Smart Art Press
Bergamot Station
2525 Michigan Avenue, Building C1
Santa Monica, California 90404
310-264-4678 (tel)
310-264-4682 (fax)
www.smartartpress.com

Smart Art Press
Volume VI, No.56

Distributed by RAM Publications
2525 Michigan Avenue, Bldg. A2
Santa Monica, California 90404
310-453-0043 (tel)
310-264-4888 (fax)
rampub@gte.net